USING INTERACTIVE DIGITAL NARRATIVE IN SCIENCE AND HEALTH EDUCATION

USING INTERACTIVE DIGITAL NARRATIVE IN SCIENCE AND HEALTH EDUCATION

BY

R. LYLE SKAINS
Bournemouth University, UK

JENNIFER A. RUDD
Swansea University, UK

CARMEN CASALIGGI
Cardiff Metropolitan University, UK

EMMA J. HAYHURST
University of South Wales, UK

RUTH HORRY
Swansea University, UK

HELEN ROSS
Helen's Place, UK

And

KATE WOODWARD
Aberystwyth University, UK

United Kingdom – North America – Japan – India
Malaysia – China

Emerald Publishing Limited
Howard House, Wagon Lane, Bingley BD16 1WA, UK

First edition 2021

Reprints and permissions service
Contact: permissions@emeraldinsight.com

British Library Cataloguing in Publication Data
A catalogue record for this book is available from the British Library

ISBN: 978-1-83909-761-4 (Print)
ISBN: 978-1-83909-760-7 (Online)
ISBN: 978-1-83909-762-1 (Epub)

ISOQAR certified Management System, awarded to Emerald for adherence to Environmental standard ISO 14001:2004.

ISOQAR
REGISTERED
Certificate Number 1985
ISO 14001

INVESTOR IN PEOPLE

CONTENTS

LIST OF FIGURES

LIST OF ABBREVIATIONS

AMR	Antimicrobial Resistance
CCE	Climate Change Education
Co-I	Co-investigator
GP	General Practitioner (medical)
h-ACE	Holistic Agentic Climate-Change Engagement Model
HEI	Higher Education Institution
IDN	Interactive Digital Narrative
IS	Infectious Storytelling
MDR-TB	Multi-drug Resistant Tuberculosis
NHS	National Health Service (UK)
NW4T	No World 4 Tomorrow
OAN	Only Always Never
PI	Primary Investigator
REF	Research Excellence Framework
Scicomm	Health and Science Communication
SEN	Special Educational Needs
SENCo	Special Educational Needs Coordinator
STEAM	Science, Technology, Engineering, Arts, Mathematics/Medicine
STEM	Science, Technology, Engineering, Mathematics/Medicine
TB	Tuberculosis
WHO	World Health Organization
WNMA	Welsh National Memorial Association
YCO2	You and CO_2

ABSTRACT

This book offers initial insights and lessons learned from two pilot studies using interactive digital narrative (IDN) as educational interventions seeking to effect positive behaviour change regarding topics of global social issues: climate change and antimicrobial resistance.

'You and CO$_2$' is a series of workshops for secondary school students: the researchers led hands-on sessions in the chemistry of carbon footprints, reading a climate-change-themed IDN and composing IDNs on the same theme. 'Infectious Storytelling' centres on affecting patient behaviours that contribute to antimicrobial resistance: in this project, researchers examine tuberculosis's (TB) representation in creative media in the Romantic era and post–World War II. This research informed a purpose-built IDN to effect positive change in public behaviour surrounding the current epidemic of antimicrobial-resistant TB (as identified by the World Health Organization).

Both these issues contribute to increasingly urgent 'global challenges': issues of climate change and ineffectiveness of medication for treatment of communicable diseases, particularly with regard to highly mobile and interspersed populations. There is a dire need to instill a stronger sense of personal responsibility to act as individuals to resolve global issues, and these pilot studies present IDNs as possible approaches in these resolutions. The studies presented in this book are an examination of the efficacy of entertainment media, specifically IDNs, to purposefully effect positive behaviour without resorting to obviously 'edutainment' games that students receive negatively.

This book's key contributions are in the areas of interdisciplinary research and education methods, combining arts and science methodologies and approaches to address significant global challenges (climate change, antimicrobial resistance). As such, it will offer insights for a rapidly growing subject area: interdisciplinary approaches. Its methodology and reflective sections on addressing the particular challenges of truly interdisciplinary research (from extremely disparate fields) will be especially helpful to future research teams.

More specifically, this book addresses science communication through interactive digital narratives. *IDNs have been shown to increase the efficacy of teaching on a range of topics, as has entertainment media in general. The IDNs at the foundation of the book's two studies were built to capture audiences' attention through strong entertainment narratives whose underlying informative and persuasive themes regarding climate change and antimicrobial resistance could affect audiences' perceptions and subsequent behaviours regarding these issues. By utilizing an interdisciplinary array of research contexts and methodological approaches, these projects aim to empirically test the effectiveness of 'playful learning' for behaviour affecting global sociological, health and environment issues; the following chapters deliver early conclusions based on the projects' pilot studies and interdisciplinary working.*

Keywords: Health Communication, Science Communication, Interactive Digital Narrative, Interdisciplinary Research, Wide Interdisciplinarity, STEAM, Entertainment for Education, Climate Change Education, Antimicrobial Resistance Education

1

INTRODUCTION

Seven researchers in seven different research areas, on two different teams, are working to approach old problems in new ways. That is the foundation of this book, which in a very short space attempts to discuss both how interactive digital narratives (IDNs) can be used for science communication (SciComm) and health communication, *and* how interdisciplinary research teams can work together. Throughout, we offer initial insights and lessons learned from two pilot studies using IDNs as educational interventions seeking to effect positive behaviour change regarding topics of global social issues: climate change and antimicrobial resistance (AMR).

You and CO_2 (YCO2) is a series of workshops for secondary school students: the researchers led hands-on sessions in the chemistry of carbon footprints, reading a climate change–themed IDN and composing IDNs on the same theme. The goal of this pilot phase of the project was to examine the feasibility of integrating 'STEAM' (science, technology, engineering, arts, mathematics/medicine) (Ge, 2015) and blended learning (PERC, 2014) approaches, specifically chemistry and IDN, to climate change education in secondary schools, specifically in the New Curriculum for Wales (Donaldson, 2015). The project's overall aim is to effect observable attitude changes in the students regarding elements of their lifestyles that contribute to climate change.

Infectious Storytelling (IS) centres on affecting patient behaviours that contribute to AMR; in this project, researchers examine tuberculosis's (TB) representation in creative media in the Romantic era and post-World War II. The former period presents TB as a romanticised affliction leading to public perception of the disease as contributing toward creative genius; the latter examines how popular film representations of effective antibiotic treatment affected public perception and behaviour. This research informed a purpose-built IDN to effect positive change in public behaviour surrounding the current epidemic of antimicrobial-resistant TB (as identified by the World Health Organization).

Both these issues contribute to increasingly urgent 'global challenges': issues of climate change and ineffectiveness of medication for treatment of communicable diseases, particularly with regard to highly mobile and interspersed populations. As issues of health and environment become global issues, the problem of *diffused responsibility* (Wegner & Schaefer, 1978) arises; the more people perceived to be responsible for resolving an issue, the less each individual feels they need to act. Despite climate change resulting from human activity, most humans feel their contribution is minimal; thus, any effort made toward reducing individual carbon footprint is futile. Likewise, individual patients feel their health is their own problem; current increases in outbreaks of formerly controllable diseases like measles show that this is not the case, and the COVID-19 pandemic has shed a harsh light on the effects individuals' actions have on collective health. There is a dire need to instil a stronger sense of personal responsibility to act as individuals to resolve global issues, and these pilot studies present IDNs as possible approaches in these resolutions.

Specifically, these studies seek to evaluate the efficacy of entertainment-based IDNs for effecting personal perception and behaviour change on these topics. As using gameplay for teaching increases, students may develop resistance to obvious 'edutainment', or media specifically created to educate (Resnick, 2004), while pure entertainment can be shown to have significantly demonstrable effects on the public – such as the television show *CSI*'s possible effect on jury behaviour (Heinrick, 2006) and celebrity endorsement of 'anti-vax' stances leading to decreases in vaccination rates and increases in outbreaks of vaccine-preventable diseases (Hoffman et al., 2017). The studies presented in this book are pilot approaches to examining how entertainment media, specifically IDNs, could purposefully effect positive behaviour without resorting to obvious 'edutainment' games that audiences receive negatively.

IDNs have been shown to increase the efficacy of teaching on a range of topics (Cai et al., 2006; Gee, 2007; Huffaker & Calvert, 2003; Mayer, 2014; Mayo, 2009; Squire, 2011b; Wright & Sandlin, 2009), as has entertainment media in general (Hoffman et al., 2017; Wright & Sandlin, 2009). The IDNs at the foundation of the book's two studies were built to capture audiences' attention through strong entertainment narratives whose underlying informative and persuasive themes regarding climate change and AMR could affect audiences' perceptions and subsequent behaviours regarding these issues. By utilising an interdisciplinary array of research contexts and methodological approaches, these projects' long-term aims are to empirically test the effectiveness of 'playful learning' (Resnick, 2004) for behaviour affecting global sociological, health and environment issues.

APPROACHES TO EDUCATING THROUGH ENTERTAINMENT

Many creatures – humans, dogs, dolphins – learn not through rote but through play (Boyd, 2009). A kitten chasing her mother's tail is learning to stalk; team-based sports have grown from tribal games honing us for war. Similarly, humans evolved narrative to convey information in oral cultures that had no other means of recording community knowledge (Boyd, 2009; Ong, 1982). If we remember advertisement jingles and film quotes better than we do the memorised facts of a lifetime of schooling, it is simply because the medium of those messages is far better suited to the way our minds retrieve and store knowledge, shaped as they were by thousands of years of oral culture versus only a few hundred years of print culture.

So, it is not really a new idea to use games and narrative for educational purposes; we just seem to have drifted away from it, pedagogically speaking. As we have (re?)discovered in the massive cultural upheaval that is 2020, while we think of schools as centres of education, they also (perhaps primarily?) fulfil purposes as childcare, socialisation and enculturation (Dorn, 2013; Gibb, 2015; Idris, Hassan, Ya'acob, Gill, & Awal 2012).[1] A common lament of artists, scholars and innovators is that today's students are taught only to take tests, to regurgitate facts and repeat interpretations as dictated by the state; they emerge from school with little capacity for creative or critical thinking, leading to a scarcity in creativity and innovation. This idea is encapsulated in Ken Robinson's extremely popular TED Talk, *Do schools kill creativity?* (2006). Rather than the lovable, rebellious icons of page and screen, who constantly work and think outside the norms of society to solve their predicaments (Han Solo, Kevin McAllister, Lisbeth Salander, River Song, Pippi Longstocking, Fox Mulder...), our structured and measured system trains our youth to conform, punishing them when they do not.[2]

---

1 In case of any doubt, we are referring to the COVID-19 pandemic/dumpster fire, compounded by astonishingly poor governmental decision-making about how to manage it in both the United Kingdom and the United States, the first author's home countries.

2 An anecdote from the YCO2 workshops: year 10s (14–15-year-olds) were the most difficult to teach, as hormones and other distractions often made them 'too cool' to participate. Dr Skains worked with a particular group of Y10s for at least 30 minutes to get them to write something – *anything* – for their Workshop 3 Twine piece; they finally responded with an absurd narrative they were certain she would denounce, and thus let them discontinue any work. In contrast to their expectations, she responded 'Great! What happens when people with eggs for heads live on the moon?' Surprised, they carried on with their story and actually had good responses to it from YCO2 team members, until their headmaster appeared and dressed them down for being 'silly', effectively stifling all that hard-earned creativity in those students. If this is what many students deal with, day in and day out, it is no wonder they shut down their own creative capacities for the sake of survival.

Pedagogical research, perhaps not unexpectedly, shows pockets of recognition of this paradigm. Rather than eschewing stories and games as simply 'entertainment' or 'time-wasting', many studies have emerged examining the health and educational benefits of narrative reading, writing and playing. As both of the IDNs for the projects in this book built on the foundations of these research insights, it is worth a brief overview of each: bibliotherapy and its counterpart, expressive writing; knowledge conveyed through mass media entertainment; the educational applications of interactive media; and persuasive communication techniques.

Bibliotherapy and expressive writing primarily fall in the reading-and-writing-for-therapy category. Bibliotherapy is a clinical therapeutic technique used as a psychological intervention for patients of all ages, using the reading of both fictional and nonfictional narratives to improve emotional states and well-being (Barker, 1995; Doll & Doll, 1997; Pardeck, 2014). Expressive writing is similar, though narrative writing (fiction and nonfiction) is used as the psychological intervention (Bolton, Field, & Thompson, 2006; Mugerwa & Holden, 2012; Pennebaker & Seagal, 1999).

We formed these projects on the foundation of health communication and SciComm, in general, and specifically how entertainment media can educate even if that is not a text's primary or intended purpose. Quite often this can include misleading or hyperbolic information; examples include the 11 April 1996 'Dangerous Foods' episode of Oprah Winfrey's talk show regarding bovine spongiform encephalopathy (BSE) infected beef that resulted in a stock market crash for beef cattle (Terry, Macy, Owens & Womble, 2016); the oft-referenced 'crime scene investigation (CSI) effect', as the popularity of so many forensic investigation TV shows gives real-life jurors false expectations about forensic evidence (Heinrick, 2006); and every positive example of a child or bystander who learned cardiopulmonary resuscitation (CPR) from TV and movies using it to save someone's life (Eisenman, Rusetski, Zohar, Avital, & Stolero, 2015).[3] As has always been the case, humans store (and use) knowledge implicit in stories, whether or not that knowledge is accurate. SciComm advocates have been pushing, as a result, for creators to take responsibility for scientific authenticity in their work, given the significant effects mass media can have on cultural knowledge and individual behaviour (Chambers & Macauley, 2015; Kirby, 2011; Kirby, Chambers, & Macauley, 2015).

---

3 The crime scene investigation (CSI) effect is largely anecdotal/mythological, as numerous studies (e.g., Alejo, 2016; Eatley, Hueston, & Price, 2016; Shelton, 2008) have illustrated. Nonetheless, the phenomenon demonstrates how popular media can shape how we think about the real world – even for prosecutors and officers of the court.

Plenty of entertainment media are known to educate their audiences, both purposefully and inadvertently. The addition of multimodal interactivity is what differentiates YCO2 and IS from other narrative knowledge conveyances, however; the focus on interactive *narrative* as opposed to gameplay likewise differentiates the projects from serious games, which are the subject of much pedagogical examination (Charsky, 2010; Cheng, Chen, Chu & Chen, 2015; de Freitas & Maharg, 2011; Flanagan, 2009; Glasemann, Kanstrup, & Ryberg, 2010). Work on multiliteracies (Cazden et al., 1996; Cope & Kalantzis, 2009a) has illustrated the benefits of multi-modal and multi-subject pedagogical techniques, combining different learning areas (reading, writing, chemistry, medicine, mathematics, programming) and offering several points of access through diverse media and communication methods (text, visual, gameplay, discussion, etc.). The YCO2 project in particular benefits from use of an IDN as opposed to games, as we enable students to not only read/play an IDN but to create one of their own; more ludologically focused strategies would steepen the coding learning curve too much for the workshop series to be feasible. For IS, the simple interactive mechanics and the web-enabled IDN are more user-friendly to the older audience who may be less practiced or interested in mobile games, as well as being a much lighter download on free public Wi-Fi or mobile data.

The project IDNs also incorporate persuasive communication and effective communicative strategies at their core. Arvind Singhal et al. (2003), in their discussions of various uses of ongoing TV series (e.g., soap operas) for health education, identify several key strategies in the specific use of entertainment narratives for education. Namely, that connection to characters is key to investing the audience in the theme of the story; that message saturation through either mass media, repetition, word of mouth or some combination of all three leads to greater communicative efficacy; that the change sought in the audience be iterative and small, so that individuals feel they can accomplish it; and that the message must comply with the audience's culture. Elements one and three (connection to characters and self-efficacy) are relatively simple to incorporate into IDNs for experienced narrative designers; tools such as narrative perspective, player personalisation, choice-based narrative and use of familiar cultural stories help to connect the audience to the characters and see how they can approach small changes in their lives. Message saturation can be more difficult, depending on who is involved on the research teams; YCO2 and IS have relied upon networking, outreach and old-fashioned hustle to enable these works and the research to gain traction. Cultural alignment, as we discuss in Chapter 2, is perhaps the most subtle and difficult aspect to accomplish. Our research teams, while crossing national, linguistic and socioeconomic boundaries, nonetheless are fairly monocultural, and our analysis shows that the result of this is a gap between our messaging and audiences that do not reflect the same cultures and attitudes. This is an element we will continue to improve, in order to reach all of our audience members.

APPROACHES TO INTERDISCIPLINARITY

In order to address such global issues involving complex issues of human need and behaviour, we assembled wide interdisciplinary (Kelly, 1996) teams. Interdisciplinary research is not necessarily a new area, and the term has become something of a buzzword for funders and institutions looking for innovative research. Research in higher education institutions (HEIs) is 'siloed', sectioned off into departments and schools by discipline; in the United Kingdom, this disciplinary partitioning begins at secondary school level, as students drop General Certificate of Secondary Education (GCSE) subjects voluntarily, and continues through to PhD level, where candidates carve out a tiny 'original' niche for themselves in a particular subject area. Yet, many real-world problems that we as researchers are seeking solutions for are complex and cannot be addressed by any one discipline alone. Thus researchers are urged, through grant opportunities and training programmes, to work across disciplines, institutions and nations to address issues like climate change and AMR, among many others.

Unfortunately, the long history of disciplinary practice in HEIs has ill-prepared them for fostering interdisciplinary research. Significant barriers to interdisciplinarity continue, including communication structures, disciplinary knowledge and language, research methods, biases about other disciplines, recognition and publication of interdisciplinary work, and even funding for interdisciplinary projects. Many researchers who would embrace interdisciplinarity have found themselves stymied, for example, by grant proposal rejections that continually place their projects *between* funding body subject areas and, thus, offer no success and no alternatives. Most academic journals are subject-specific, making publication of cross-disciplinary work difficult, and measures of research quality such as the UK's Research Excellence Framework (REF) are judged according to disciplines and disciplinary standards. When academics are already overwhelmed by increasing casualisation and teaching/admin workloads, it can be an insurmountable task to blaze new pathways through the venerable mazes of historical HEI infrastructure for interdisciplinary research.

Yet that is exactly what we have achieved on the projects discussed in this book. As recounted in Chapter 5, these projects were enabled by a research leadership programme that actively sought adventuring researchers and trained them to communicate and work across disciplinary boundaries. In her foundational text on interdisciplinary research, Julie Thompson Klein (1990) notes that few accounts of interdisciplinary work exist, and far fewer of wide interdisciplinary, or arts-science, research. Most discourse on interdisciplinarity, in fact, centres on research that incorporates multiple *science* disciplines and fails to address arts-science research altogether. Indeed, many so-called arts-science interdisciplinary projects are not truly interdisciplinary as we have defined it (combining multiple disciplines to approach a complex problem); rather, they incorporate one subject in servitude to another, such as a textile artist who incorporates replicated DNA into her fabrics or a scientist who writes popular fiction to convey fundamental concepts to wider audiences. These endeavours are laudable and fascinating but are distinct from interdisciplinary research as we have applied it on these projects. Thus, we hope to provide one of those rare inside looks into wide interdisciplinary teams, demonstrating how we overcome the infrastructural barriers to this type of work, and how others can do the same.

THE AUTHORS

Given that the projects and work explored in this book originate from two unique teams across disciplines, we include a brief biography of each researcher here. Each co-investigator has contributed to the ideation, design, implementation, analysis and outcomes of the YCO2 and Infectious Storytelling projects, albeit in different ways according to their areas of expertise and personal skill sets. Without knowing our backgrounds and interests, it would be difficult to parse exactly how crucial and influential each of us are on these projects; thus, we have allowed ourselves the luxury of introducing ourselves in text.

R. Lyle Skains

Primary Investigator, Infectious Storytelling
Co-Investigator, You and CO_2

Lyle researches and teaches Creative Writing and Digital Media, exploring multimodal creativity, genre fiction, writing and reading/playing transmedia narratives, and writing and publishing in the 21st century. Her research is largely practice-based, stemming from her work in creative writing (speculative fiction) and digital writing; her PhD explored how digital writing practice affects writer and narrative. Prior to her career as a writer, she studied to be a biological anthropologist, studying evolutionary genetics and primate morphology. In her practice-based research and teaching, she builds upon her experience as a professional writer in prose and the film and technical industries. Lyle is interested in exploring the ways that digital media, fiction in particular, can be integrated into our everyday lives, including education, entertainment, communication about specialist topics (science, policy, health, culture, etc.) and personal exploration.

Her specialised work on YCO_2 and IS is to design and create the IDNs for both projects.

Jennifer A. Rudd

Primary Investigator, You and CO_2

Jennifer is a slowly reforming technical scientist. After almost a decade researching technological solutions to climate change on two continents, she had an epiphany in 2018 and realised that the communication of climate change and its solutions were far more important than any lab-based advance she could hope to make. Therefore, with a background in the chemistry of solar panels, water splitting for a hydrogen economy and converting carbon dioxide into the fuel of the future, she turned her attention to climate change education. Jennifer writes climate change education material to suit a variety of ages and uses her communication skills to convey the severity of the climate emergency through national talks, radio and printed media. Besides being the primary investigator of YCO2, Jennifer is the programme manager of the recently funded Circular Economy Innovation Communities project based at Swansea University, which implements 'circular economy thinking into public services [to] help reduce carbon footprint and [form] part of the solution to the global climate emergency' (Newman, 2020, n.p.).

Her specialised work on the YCO2 project is in developing teaching materials for the carbon footprint education component, recruiting schools to participate and engaging in outreach work.

Carmen Casaliggi

Co-Investigator, Infectious Storytelling

Carmen completed her undergraduate degree at the University of Milan and her PhD at the University of Kent. She is now working as a Reader in English literature at Cardiff Metropolitan University with research and teaching expertise in the Romantic period. Her most current research focuses on the transdisciplinary legacies of Romanticism in the arts and humanities, with particular reference to the construction of concepts of transnational sociability, nationhood, identity and politics of migration, alongside present-day debates on global environmentalism.

Carmen is keen to explore ways to use Romantic literature in applied settings, including issues such as climate change and environmental sustainability and the dialectic between capital and globalisation. She has enthusiastically embraced the exploration of TB in the Romantic period and how it influenced art and literature of the age.

Her specialised work on the Infectious Storytelling project centres on researching TB's representations in the Romantic era, specifically with regard to representations that influenced treatments and patient behaviour.

Emma Hayhurst

Co-Investigator, Infectious Storytelling

Emma obtained her BSc degree in Microbiology at the University of Sheffield. She then stayed on to study for her PhD on bacterial cell wall dynamics. After two years of postdoctoral research, she moved to South Wales in 2009 to take up her current position as Senior Lecturer in Microbiology at the University of South Wales. In 2016, Emma was awarded a National Research Network for Low Carbon Energy and Environment (NRN-LCEE) Returning Fellowship, and she has been actively involved in applied research in the area of antibiotic resistance since then.

Her research interests are broad and varied, and funded projects include: working in collaboration with Welsh Water to assess the prevalence of antibiotic resistance throughout wastewater treatment plants; the development of a low-cost device for the specific detection of *E. coli*; and working in collaboration with the National Health Service to assess the infection risk associated with the use of technology in clinical settings.

Emma's specialised work on IS is in AMR, including its mechanisms and effects; she expanded her research area to include medical history with regard to TB and antibiotics.

Ruth Horry

Co-Investigator, You and CO_2

Ruth obtained her BSc degree in Psychology at the University of Nottingham, before completing her Master's Degree in Research Methods and PhD at the University of Sussex. Following her PhD, Ruth worked as a postdoctoral researcher at Royal Holloway, University of London, from 2009 to 2010, and then at Flinders University in South Australia, from 2010 to 2014. In 2014, Ruth took up a lectureship in Psychology at Swansea University, where she is now a Senior Lecturer.

Ruth is a cognitive psychologist who specialises in applying our understanding of human memory and decision-making to real-world contexts. Primarily, her research has focused on the legal context, examining how witnesses recall details of crimes they have witnessed, how accurately witnesses can identify a previously seen perpetrator and how jurors reach verdicts in criminal trials.

Within the YCO2 project, Ruth has overseen the development of survey measures designed to evaluate the effectiveness of the programme in influencing understanding of, and attitudes towards, carbon footprint reducing behaviours.

Helen Ross

Co-Investigator, You and CO_2

Helen is a Special Educational Needs (SEN) specialist and dyslexia expert. She is passionate about supporting young people to be their best 'self' through holistic intervention, with a focus on building positive working relationships with all stakeholders in their education. Helen is a believer in evidence-based practice being a vital tool in teaching. She is an active researcher-practitioner, having received her PhD on the experiences of young people with dyslexia, who promotes practitioners engaging with research and vice versa through her paid work and her voluntary positions.

Helen started teaching in Barnsley, South Yorkshire, and worked initially as a secondary maths and French teacher. She is a fully qualified Special Educational Needs Coordinator (SENCo), and alongside providing support to other professionals and undertaking research, she currently works part-time as a SEN teacher in a mainstream school. In addition to her paid work, Helen also undertakes voluntary roles: at a local primary school, she provides critical oversight of special needs provision within the school, works alongside the SENCo to sure statutory obligations are met effectively and provides advice relating to 'best practice'; she also volunteers there one morning per week, where she supports learners who find literacy difficult. Helen is Chair of the Wiltshire Dyslexia Association, where she supports the running of events, provides expert advice on pedagogy and contributes to the Association social media networks.

Helen's specialised work on YCO2 is consulting on the creation of differentiated project materials for different learning needs, including supporting materials for teachers and pupils, and qualitative analysis of pupils' original IDNs and focus group data.

Kate Woodward

Co-Investigator, Infectious Storytelling

After periods working in the television industry and the arts sector, Kate returned to the Department of Theatre, Film and Television Studies at Aberystwyth University to complete her PhD. She is now a Lecturer in Film Studies at the same institution where she teaches through the medium of both Welsh and English. Her current research interests focus on screen fictions, particularly the way that landscape, place and location are represented in and are sometimes transformed by film and television dramas.

Kate has also held some public positions, including being Vice-Chair of the Arts Council of Wales between 2012 and 2017. She is keen to develop public outreach activities to communicate her research to all and strategies for her research to benefit Wales.

Her specialised work on Infectious Storytelling is to analyse the representation of TB and antibiotic use in film, particularly in UK settings and post-World War II films, and to analyse the media campaigns from national and international bodies such as Public Health England and the World Health Organization.

OVERVIEW OF CHAPTERS

The following chapters offer initial results on the interdisciplinary research projects YCO2 and Infectious Storytelling, followed by insights on the creation of their IDNs and the workings of the teams.

Chapter 2, Pilot Case Study: YCO2, provides an overview of this climate change education project's pilot phase from methods to current outcomes. The chapter focuses on the initial ethnographic, experiential and reflective data from the researchers, primarily with regard to the in-classroom workshops. It also includes an extended discussion based on qualitative, Bourdieusien analysis of the participants' submitted IDNs on climate change, indicating that the first iteration of the project IDN (*No World 4 Tomorrow*) and workshops contains gaps in reaching students of differing cultures and socioeconomic backgrounds.

Chapter 3, Pilot Case Study: Infectious Storytelling, similarly offers an overview of the project, including an extensive discussion of its background research into representations of TB in media in the Romantic and post-WW2 eras. The chapter includes a description of its methodological approach and observations on the pilot phase of focus group studies. This initial phase indicates that a segment of the target audience for the project IDN, *Only Always Never*, has strong affinity for the story and responds positively to its message. We also discuss future stages of this project, including revisions and longer-term observation of the effects of the IDN on patient behaviour.

The lead author, Lyle Skains, takes Chapter 4 (Entertaining to Educate: Creative and Pedagogical Insights) to discuss the insights into creating IDNs for SciComm, developed through practice-based research. She explores the supporting research for using IDNs for these purposes and describes her approach to creating both *No World 4 Tomorrow* and *Only Always Never*. She particularly focuses on the differences between designing interactive narratives for entertainment purposes versus for educational purposes and how educational aims affect the writer and the narrative. As few to no texts exist to guide writers in creating IDNs specifically for SciComm, this chapter offers unique perspective and guidance on this process.

Chapter 5, Bridging Research Silos: Approaches to Arts-Science Collaboration, broadens once again to encompass all co-investigators on these projects as it examines the ins and outs of interdisciplinary research. The chapter begins by defining interdisciplinarity and discussing the history of interdisciplinary research in the disciplinary world of higher education. We review the barriers to interdisciplinary working, as well as the benefits, and analyse the project teams for their processes of conducting these projects together from disparate disciplines and perspectives.

These perspectives are highlighted further in Chapter 6, Lessons Learned: Researcher Reflections. Each co-investigator contributed a reflective commentary on their work on YCO2 and Infectious Storytelling, focussing on their experiences, challenges, contributions and responses to these interdisciplinary teams and projects. We first contextualise these accounts in comparison with Chapter 5's recommendations for successful interdisciplinary work, then offer the commentaries in full, as we came to understand that they are most valuable to other researchers examining interdisciplinary research in their original unedited forms.

The final chapter draws the book to a close, summarising our findings on the pilot projects and our insights into creating IDNs for SciComm and working in interdisciplinary teams. Most importantly in this concluding chapter, we look forward to the next steps for these projects, our planned expansions and expected outcomes.

The projects outlined in this book have had ups and downs, periods of intense activity followed by stretches of utter silence. They have been deeply affected, as has all work and relationships across all sectors and locales, by the COVID-19 pandemic. Nonetheless, we collectively consider these projects successes in their initial stages, and every member of each team has recommitted to their roles on the projects, eagerly looking forward to the next phase of challenges and insights. We hope that our work as presented here can provide a strong foundation for wide interdisciplinary research across the arts and sciences, and offer a model for implementing IDNs for health communication and SciComm.

2

PILOT CASE STUDY: YOU AND CO$_2$*

INTRODUCTION

The project described in this chapter, You and CO$_2$, originated with a central question: how do we convince teenagers to not only change their opinions about climate change but also change their *actions* related to climate change? This global challenge is urgent and timely, but also overwhelming. To resolve it will take active, positive change from every corner of society, from the individual to industry and government – change that has prompted resistance rather than action. Climate change education (CCE) efforts often focus on conveyance of information and appeals to emotion (in 2018, when we brainstormed the project, school strikes and Extinction Rebellion were not yet international phenomena). We theorised there may be a fundamental element of psychology at play: that the notion of distributed responsibility (Wegner & Schaefer, 1978) was diluting individuals' sense of personal responsibility so much that they failed to take action even if they agreed that climate change was an urgent matter. If the fault for global catastrophe rests equally on 7 billion different shoulders, what is the use of one individual?

We approached the You and CO$_2$ (YCO2) programme from four inter-disciplinary perspectives: chemistry (carbon emissions/carbon capture), psychology (effecting and measuring behaviour change), pedagogy (innovations in

* Elements of this chapter were originally published as: Rudd, J.A., Horry, R., and Skains, R.L. (2019). You and CO2: a Public Engagement Study to Engage Secondary School Students with the Issue of Climate Change. *Journal of Science Education and Technology.* [online]. Retrieved from http://link.springer.com/10.1007/s10956-019-09808-5 [Accessed on March 3, 2020]; and Ross, H., Rudd, J.A., Skains, R.L., and Horry, R. (2021). How Big is My Carbon Footprint? Understanding Young People's Engagement with Climate Change Education. *Sustainability,* 13(1961). Retrieved from https://doi.org/10.3390/su13041961 [Accessed on February 11, 2021].

science, technology, engineering and mathematics [STEM] and arts crossover models) and electronic literature (creating and experiencing interactive fictions). Our goal was to design an element of secondary school curriculum (specifically in Wales, initially) that communicated the science of climate change while engaging the students in multiple literacies, including reading, writing, multimodal communication, programming, working in groups, creativity, narrative and nonlinear thinking. Together we designed a model programme and trialled it in two secondary schools in Wales; this chapter presents our initial findings and experiences. Below, we outline our rationale for the approach before describing the programme and evaluation methods more fully.

Urgency of School-based Climate Change Engagement

According to the 2018 Intergovernmental Panel on Climate Change special report, the world has already warmed by 1.1 degrees centigrade compared to pre-industrial levels (Mora et al., 2017). Should warming continue at this rate, changes to the natural world will significantly affect the habitability of the planet for humans (ibid.). Global warming can only be limited by reducing global greenhouse gas emissions (IPCC, 2014). In the United Kingdom, *Carbon Brief* estimates that on average individual citizens contribute 5.4 metric tonnes of CO_2 (Evans, 2019, n.p.); Dietz et al. estimate that household carbon emissions can be reduced by 5–12% if householders adopt a range of behavioural changes (e.g., line drying clothing, reducing thermostat settings) (2009). Other activities produce much less obvious, indirect emissions arising from the production of goods and services (Berners-Lee, 2008; Wiedmann & Minx, 2007); choices we make regarding diet, clothing and entertainment (such as Internet use) also carry associated carbon emissions (Gombiner, 2011; Kim & Neff, 2009).

Individuals' contributions to climate change pale in comparison to industry and government (Griffin, 2017; Heede, 2019a; 2019b), which is also a contributing factor to inaction, as many feel their positive actions are ineffective when compared to the emissions produced by corporations. Nation-states do have a significant role to play in the effort to reduce emissions, as formally acknowledged in the signing of the 2015 Paris Agreement. Political power, however, has favoured campaign contributions and economic benefits for corporations rather than individual voters (Beder, 2014); until significant elections and, thus, political will hinge on the issue of climate change, it is unlikely that governments will force climate catastrophe's primary contributors to alter their practices. Thus, not only does CCE need to inspire people to act to reduce their own personal carbon footprint, it also needs to persuade them to make the climate a priority in terms of voting, contributing to charitable causes and participating in activism.

For individuals to understand this need for attitude and behaviour change, education and engagement with climate change science is key; K–12 students with more education in climate science express increased engagement in climate change-related activism than others (Lester, Ma, Lee & Lambert, 2006). While individuals of all ages would benefit from increased engagement with climate science, there are several reasons to prioritise young people in CCE programmes. First, the youngest citizens of our world will have to live with the consequences of climate change for longest and will, therefore, be the most heavily impacted by those consequences (Jorgenson, Stephens, & White, 2019). This fact is increasingly recognised by children and adolescents them-selves, leading to unprecedented levels of activism from young people; indeed, since 2019, young people have played a major role in raising global awareness of climate change and its consequences. The School Strike movement, for example, has grown from one Swedish teenager to an estimated 1.4 million students globally as of May 2019 (Evensen, 2019) and 7.6 million people as a whole as of September 2019 (350.org, 2019).

Second, behaviour change becomes increasingly difficult as habits become more firmly entrenched (Webb, Sheeran, & Luszczynska, 2009). Conse-quently, many attempts to change behaviour across a broad range of domains (e.g., diet, exercise, smoking) fail completely or end in relapse (Polivy & Herman, 2002). Adolescents are less likely to have formed many of the habits that contribute heavily to household carbon emissions (Dietz, Gardner, Gilligan, Stern, & Vandenbergh, 2009); this provides the opportunity to intervene before they form carbon-intensive habits.

Third, the education system provides a unique environment for public engagement with science. The nature of schooling means that it is much more practical to develop programmes that span multiple sessions than it is for typical science outreach and public engagement activities, which tend to be delivered as standalone events. Programmes delivered in a school environment can, therefore, be more ambitious, providing more scope for combining multiple approaches in an effort to increase engagement.

The current political and social environment also makes the current programme, and others like it, timely. In the United Kingdom, there is considerable appetite among teachers and students alike for greater coverage of climate change in the school curriculum (Taylor, 2019). At present, climate change tends to be covered in individual subjects such as science and geography, though there are recent calls for climate change to be integrated as a core theme throughout the curriculum (Rayner, 2019). Furthermore, a recent review of the education system within Wales argued that a fundamental goal of education should be to develop 'ethical, informed citizens who… understand and consider the impact of their actions when making choices and acting' (Donaldson, 2015, p. 30). The You and CO_2 programme was developed to allow students to explore ideas about their own role as a consumer and a member of society in tackling the urgent global issue of climate change.

Climate Change Education Framework

Recent CCE literature indicates that effective educational programmes should cross traditional disciplinary boundaries and encourage students to reflect on the broader social and moral context of climate change (cf. Ardoin, Bowers, Roth & Holthuis, 2018; Cantell, Tolppanen, Aarnio-Linnanvuori & Lehtonen, 2019; Gayford, 2002; Jorgenson et al., 2019; Pruneau et al., 2001; Wise, 2010). Teachers, however, tend to separate topics into silos, finding it easier to 'maintain the integrity of their subject rather than be involved in extensive interdisciplinary teaching' (Gayford, 2002, p. 1191). Climate change, however, is a multidisciplinary problem with a need for multidisciplinary solutions. We, therefore, developed our programme using combined science and arts-based approaches or STE(A)M (science, technology, engineering, [arts], mathematics/ medicine), incorporating arts and humanities into science-based teaching.

A significant proportion of CCE research supports and recommends inter-disciplinary and multiliteracies approaches, as climate change is an interdisciplinary problem stretching across chemistry, geography, social science, politics, economics, psychology, health and more. Nicole M. Ardoin et al.'s 2018 work demonstrates that CCE can be used to develop the following literacies: *systems thinking* (holistic approaches to analysis), *critical thinking* (objectively evaluating facts to form judgements), *decision-making research skills* (synthesising data into knowledge for recommendations and action) and *science-process skills* (observing, communicating, classifying, inferring, measuring and predicting). Likewise, Diane Pruneau et al. (2001) and Sarah B. Wise (2010) both recommend an interdisciplinary approach to CCE; Wise in particular recommends supporting the teachers through professional development activities, specifically regarding interdisciplinarity. She suggests that science and social studies departments could work together to teach climate change, since 'disciplinary divisions… appear to generate barriers to providing students with comprehensive instruction about climate change' (p. 305).

In addition, the need to stimulate hope for the future echoes throughout CCE studies. Pruneau et al. (2001) recommend 'future education' where students are encouraged to re-imagine their future; by doing so, students are empowered to imagine changes in their lives and explore the consequences of those changes. Pruneau et al. also recommend a critical socio-constructivism approach where CCE is 'presented as a generalized discussion' (p. 134) wherein instructor and students engage in a group chat on the topic. Simon N. Jorgenson's group (2019) recommends that CCE focus on 'local, tangible and actionable' (p. 165) endeavours that can be achieved by individuals. This would then inspire inter-generational education where the student would bring home CCE from school and, therefore, influence the behaviours of their caregivers. Their recommendations are to:

(1) Move CCE beyond a focus on individual behavioural change; teach about systemic change.

(2) Develop participatory CCE to engage students in multi-actor networks of NGOs, climate scientists, community groups, state agencies and renewable energy firms.

(3) Teach about technological and social innovation currently occurring in the world; this engenders hope, which in turn leads to meaningful action.

(4) Develop new narratives to encourage long-term engagement with CCE.

Hannele Cantell et al. (2019) extensively reviewed CCE resources and developed the Bicycle Model of CCE. The model summarises the key components of CCE, using the image of a bicycle as a visual representation of those components: a frame consisting of *values*, *world view* and *identity*; pedals providing *action* impulse to the wheels of *knowledge* and *thinking skills*; a saddle of *motivation and participation*; *operational barriers* braking the system; handlebars steering toward a *future orientation* as *hope* and *emotions* lighting the road ahead. Cantell et al.'s model also makes the following key recommendations to those developing new CCE materials:

(1) Emphasise that humans can change society; engage students in joint positive action.

(2) Encourage students to think about the role of human beings as consumers and, therefore, the cause of environmental problems.

(3) Stimulate hope and compassion.

(4) Combine science-based teaching with critical thinking, so that students can assess technological advances within a broader social context.

This model, thus, creates a holistic, interdisciplinary approach to CCE that engages students both in developing skills across an array of STEAM subjects while also contextualising climate change in a broader sociological sense. As such, we have used these recommendations to evaluate the efficacy of our workshops (see the Discussion section below).

Engaging Students through Arts-based Approaches

The You and CO_2 project adopted a multiliteracies approach (Cope & Kalantzis, 2009; Skains, 2019a), incorporating digital literacy, interactivity, creative writing, game design, discussion, and group and individual work to engage students with the topic of climate change. The first workshop focused on the chemistry of climate change and provided the opportunity for students to explore their own carbon footprints. The second and third workshops incorporated bibliotherapy with an interactive digital narrative (IDN) and expressive writing through game design and coding to allow the students to explore ideas around personal responsibility, climate change and the consequences of climate change in a novel and engaging way. Bibliotherapy employs purposeful reading as a psychological intervention for treatment of clinical issues (Pardeck, 2014); here, we apply it to the more technological medium of interactive, hypertext, digital narratives or 'Twine games' (Klimas, 2009). Likewise, expressive writing uses writing narratives as a therapeutic intervention in cognitive behavioural therapy in order to address emotional or mental issues (Bolton, Field, & Thompson, 2006; Mugerwa & Holden, 2012; Pennebaker, 1997); in this programme, we adapt this concept to computer-based interactive narrative design. Both bibliotherapy and expressive writing have the potential for broader psychological interventions; here, we use them as a means to engage students more deeply with the core themes of the You and CO_2 project—namely, their own role as a citizen of the world in limiting carbon emissions.

To increase the immersiveness of the bibliotherapy and expressive writing components of the You andand CO_2 programme, we used the technology of *digital fiction* or IDN: fiction that is written specifically to be read from a digital device (e.g., computer, tablet). IDNs make full use of the digital environment to incorporate elements (e.g., branching plot lines, moving images) that would be impossible in an analogue format (see Bell et al., 2010 for a full definition). The use of IDNs in bibliotherapy and expressive writing is relatively new, so evaluation is limited (although see Ensslin et al. 2016 for a study on digital fiction/IDNs as tools for teenage body image bibliotherapy).

However, our working hypothesis is that playing and coding IDNs will embed the core themes from the You and CO$_2$ project on multiple cognitive levels, thus creating deeper engagement with the themes covered in the programme.

METHODS

In consultation with secondary school teachers, we designed an interdisciplinary series of workshops that could be embedded in secondary school curricula, aligning with the principles of the New Curriculum for Wales (Donaldson, 2015). Lyle Skains created a purpose-built IDN for the project, *No World 4 Tomorrow*, the process of which is discussed in Chapter 4. This section outlines the pilot schools and student groups in our sample set, the workshops and our data analysis approach.

Sample

Secondary school students from two schools in Wales, UK, participated in the workshops. In School 1, 85 students from three Year 9 classes (ages 13–14) participated. School 1 was a large comprehensive school in South Wales with around 2,000 students from 11 to 18 years of age. The school serves a relatively affluent area, with a below-average proportion of students eligible for free school meals. The proportion of students with special educational needs is below average for Wales. Around 25% of students are from a minority ethnic background (Estyn, 2018).

Ninety-five students from School 2 participated. The students were in Years 8, 9 and 10 (ages 12–15). School 2 was an independent school in North Wales with around 200 students aged 9–18 years. As a fee-paying school, most students come from relatively affluent backgrounds. While the majority of students live in the surrounding areas, around one-third of the students live internationally and board at the school during term time. Very few of the students have special educational needs. Around 23% of the students are from minority ethnic backgrounds, with around one-third of students speaking English as an additional language (Estyn, 2018).

To protect student anonymity, we did not record the gender or age of participants; however, the gender balance across each of the classes was approximately even.

Workshops

The programme involved three workshops delivered by the research team. In School 1, the workshops were each roughly one month apart. In School 2, the workshops took place in much more rapid succession, with the entire programme delivered over three days. Each workshop is described fully below. Briefly, Workshop 1 focused on the chemistry of climate change and the carbon emissions associated with everyday activities that were relevant to the students' own lives. Workshops 2 and 3 introduced the IDN component of the programme. In Workshop 2, students played through a custom-written IDN, *No World 4 Tomorrow* (*NW4T*), while in Workshop 3, students created their own IDNs on the theme of climate change.

Workshop 1

Workshop 1 was designed to ensure that students had a basic understanding of the role of carbon dioxide (CO_2) in climate change and to encourage students to reflect on the carbon footprints associated with their everyday activities. After reflecting on our experiences running Workshop 1 at School 1, we made several changes to the session before delivering it at School 2. Below, we focus on the revised workshop delivered in School 2, highlighting key points of difference for the sessions in School 1. Lesson plans for Schools 1 and 2 can be found on the You and CO_2 website.

Workshop 1 began with group discussions probing students' understanding of the term 'carbon footprint'. Students volunteered types of human activity (e.g., energy, transport) that contribute to carbon footprints. In the next part of the session, we aimed to bring to life the process by which CO_2 is created through hands-on activities. Alongside instruction about the chemical reactions involved (see Eq. 2.1), students created three-dimensional models of methane and oxygen molecules using Bunchems (Velcro-style balls of different colours that can be stuck together), which they converted into CO_2 and water molecules (see Fig. 2.1).

$$CH_4 + 2O_2 \rightarrow Energy + CO_2 + H_2O \qquad (2.1)$$

The remainder of the session focused on the carbon footprints associated with everyday activities (e.g., travelling by different modes of transport, consuming different breakfast foods) that would be relevant to the students' own lives. Carbon footprints for each activity were taken from the book *How Bad Are Bananas* (Berners-Lee, 2008)? and the Tesco supermarket website. To help the students visualise the carbon emissions associated with each activity,

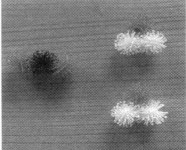

Fig. 2.1. Bunchems Models of the Reaction by Which Methane and Oxygen Combine to Form CO$_2$ and Water.

they were shown an inflated balloon, which they were instructed represented 16g of CO$_2$. The number of balloons was then used as units for the remaining activities and discussions; for example, travelling one mile in a car was described as equivalent to 44 balloons of CO$_2$.

In School 2, students calculated their own carbon footprints for the first two hours of their day using the reference document in the online supplementary materials. Time allowing, the students were then grouped together and asked to reduce the group's carbon footprint by one-third. In School 1, as a whole class activity, students guessed the number of balloons of CO$_2$ associated with a number of different activities.

Workshop 2

In Workshop 2, we aimed to encourage students to reflect on the importance of living in an environmentally sensitive way through immersion in an interactive story entitled *NW4T*, which was written specifically for this project by Lyle Skains (freely available at www.youandco2.org; see Fig. 2.2 for a structural map of a sample section). Throughout the story, readers are able to make choices about how the characters behave—from choosing the food that they consume, to how they travel, to the way in which they engage with their community concerning societal issues around sustainability. The actions that the reader makes affect the direction of the story, ultimately leading to one of six possible endings. These endings include being passive and letting disaster happen, accepting personal responsibility but eschewing dramatic action, actively engaging in *accelerating* the oncoming disaster, engaging and becoming an eco-warrior, profiting from the oncoming disaster, and becoming an activist, thus saving the world. Additional nuance was given to each ending

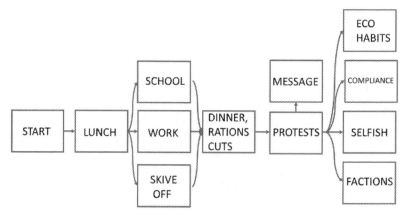

Fig. 2.2. A Simplified Excerpt of the Backend Structure of No World 4 Tomorrow, Showing the Multiple Pathways That the Interactive Digital Narrative (IDN) Takes Dependent on Choices Made.

depending on choices made throughout the story, indicating how small, everyday choices can lead to varying levels of personal impact on global environment.

The lesson plan for Workshop 2 can be found on the You and CO_2 website. At the beginning of the workshop, students were introduced to the concept of interactive narratives by looking through a number of published examples. Digital fiction, or IDN, was defined as 'fictional stories created ON or FOR digital devices, which would lose important elements if taken out of digital media' (Bell et al., 2010, n.p.). The students then played through the story as a whole class. Students made collective decisions about characters' names and actions. For some decisions, where there was a clear majority for one option, that option was chosen. For other decisions, where there was no clear consensus, the students were encouraged to debate the possibilities until they could agree on a decision.

Following the whole class play through, the students worked in small groups to discuss the following questions: How does this story relate to the topics covered in Workshop 1? How similar are the characters' experiences and choices to your own experiences and choices? How can you relate this story to your own experiences and habits?

In the remaining time, each student was given individual access to the IDN to play through the story by themselves or in pairs. They were encouraged to find as many different endings as they could by making different decisions for the player character.

Finally, as homework, students were asked to prepare for Workshop 3 by planning a storyline for their own IDN. They were encouraged to think about details such as the kind of world their story would be set in; whether it would be set in the past, present or future; who the characters would be; and what would happen, including the possibility for multiple plot lines.

Workshop 3

The aim of Workshop 3 was for students to develop their own ideas by creating their own IDNs. We took a deliberately non-prescriptive approach, allowing the students to focus on whatever themes they chose (as long as it related to climate change and/or carbon footprints). Students were introduced to Twine, an open-source programme for digital storytelling (https://twinery.org) and were directed to online tutorials housed on the project website. These tutorials begin with accessing the programme and creating a new hypertext project, getting familiar with the architecture of the software and learning the fundamentals of coding for digital functionality. Students desiring added functionality (such as custom colours, user input, images, sound, points systems, etc.) can work their way through the more advanced tutorials (also created by Skains) that incorporate Hypertext Markup Language (HTML), JavaScript and Cascading Style Sheets (CSS) coding. In School 2, where more time was allocated to the session, Skains gave the students additional in-person tutorials. Following this introductory segment, Workshop 3 was relatively unstructured, with students free to work on their IDNs at their own pace, while members of the research team circulated the room to answer queries and provide support.

Evaluation

We evaluated the programme in two key ways.[1] First, we reflected on our own experiences of administering the programme, which was facilitated through structured conversations with teachers in School 1. We reflected on both

---

1 We also pilot tested a self-report scale designed to measure participants' attitudes towards reducing their own carbon footprints. However, in the absence of a control group, the data are not readily interpretable. We, therefore, refrain from presenting these data here, though the scale, and data are openly available on the project's Open Science Framework page: https://osf.io/ w874b/.

technological and logistical challenges with delivering the workshops, as well as the extent to which the students appeared to be engaged with the content.

Second, we applied qualitative analysis to the participants' original IDNs. Students were free to submit or decline to submit their own work to the project website. In total, 85 stories were submitted (55 from School 1; 30 from School 2); of which, 79 were suitable for qualitative analysis. This qualitative analysis was constructed from a grounded theory perspective (Glaser & Strauss, 1967) in four iterations, beginning with broad 'core categories' (Glaser, 1978) and progressing until the primary analyst (Ross) identified six types of student responses to the programme as illustrated by their IDNs (discussed below).

Given structural limitations on individual action and capacity to engage with CCE (depending on their cultural norms and value systems), we under-pinned the qualitative data analysis of the student IDNs with a Bourdieusien framework, which situates students' capacity to engage with a particular topic according to the cultural expectations (habitus) and value systems within the field of education (Bourdieu, 1977).[2] Bourdieu's sociology is an 'almost self-conscious intervention' that explores inter-agentic social relationships within a field and delineates factors affecting actor/group responses to structures (Barnard, 1990, p. 62). The key feature of Bourdieu's sociology is its consideration of both the observed 'subject' (i.e., student) and the structures affecting their reactions (i.e., social class, curriculum), how they are inculcated towards certain behaviours and how habitus informs value systems of their position within any social field (Barnard, 1990; Bourdieu, 1977). As a core aim of the You and CO_2 project is to inspire young people to individual action and structural change regarding the climate crisis, this sociological perspective is, thus, well-suited to frame the analysis.

RESULTS

On the whole, the workshops were well received by the students and the teachers. Here, we reflect on the logistical and situational challenges that we faced and how these informed the steps we took to refine the workshops. We also outline improvements that we have identified for the programme going forward and highlight issues that readers may wish to consider if adapting the You and CO_2 programme, or using a similar approach, for implementation.

2 We note engagement capacity may also be affected by curricular content (Competente, 2019) and/or good/poor pedagogical practice (McNeal & Petcovic, 2019).

Reflections on Programme Delivery

In School 1, large class sizes (around 30–35 students) presented a challenge to maintaining order; waning student attention often resulted in widespread chatter. Workshop 1's first instance (in School 1) was instructor-led, with minimal peer-to-peer interaction. The researcher-instructor primarily interacted with the students in a Q&A format, seeking student responses on a hands-up basis; this resulted in engaging only the most outgoing and confident students while other students' attention drifted.

We revised Workshop 1 based on these reflections, as well as focus group discussions with School 1 teachers. In its second instance (in School 2), we refined the session to increase interactivity and peer-to-peer interaction, creating a more engaging and inclusive workshop that worked much more effectively. We introduced a group work component in which students calculated lifestyle changes to reduce their group's overall carbon footprint by one-third; this exercise helped students identify sources of their biggest CO_2 emissions. It also provided valuable role-playing experience for carbon reduction negotiations, mirroring, albeit in a simplified way, the conversations that will happen in governments as they move towards legislating for a low-carbon economy.

We began Workshop 2 by reading through the IDN as a whole class activity. We did so to guarantee that all participating students would experience the story in full at least once. We found, however, that class size significantly altered the efficacy of the group reading. In smaller classes (15–25 students), the students enjoyed the somewhat chaotic but collaborative nature of making group decisions and debating the available story options. In larger classes, however, this approach led to general disruption and difficulty in guiding the class through the IDN. Teacher focus groups for these classes (School 1) suggested individual playthroughs for Workshop 2 as opposed to the group reading; we will trial this approach in future iterations of the project.

Workshop 3 provided us with an informal insight into the impact of preparation on programme delivery, based on the workshop timescales between the two schools: School 1 workshops were spread over three months, while School 2 workshops occurred during a 'STEM Week' on three concurrent days. As such, School 1 students were able to plan out their own IDNs as 'homework' between Workshops 1 and 2 (though only one class of three had done so); due to the tight turnaround between Workshops 2 and 3 at School 2, homework was not feasible. Overall, we found that the session in which the students came prepared with story ideas and even some plotting maps worked more effectively, as students were able to make the most of their time, therefore producing more complete IDNs than in the other groups. Indeed, in classes that had not prepared ahead of the session, the most frequent query from students was along the lines of 'I don't know what to write about'. Future instances of the You and CO_2 programme should incorporate either homework and/or additional planning sessions.

Workshops 2 and 3 required technical equipment, including access to computers with Internet connectivity. We designed these workshops for the ideal 1:1 ratio of student to computer; consistent technical and logistical challenges, however, resulted in several students typically sharing one device. Slow Internet speeds also caused significant issues in School 1, which limited the ability of some students to play through the story individually in Workshop 2. We also found that the schools' firewalls blocked the project website and thus the IDN, creating delays in beginning Workshop 2 for the first classes in both schools. Future instances of the You and CO_2 workshops plan to liaise closely with the technical staff to discuss bandwidth capabilities and computer access in advance of the session, as well as ensuring that any required websites are unblocked. As these technical issues not only increased time pressure in the workshops but also created discipline issues as students grew impatient, it is important to arrange necessary materials, devices and full access in advance of all workshop sessions.

On a technical level, none of the students struggled using Twine. Some students were more adventurous than others, while others created more traditional, linear stories. The former incorporated customised elements such as colours, user input boxes, images and sound files; all students incorporated at least the basic level of Twine coding for links between passages and some aspect of choice-based branching structures. Informally, we observed that younger students (Year 8, ages 12–13) tended to use the software quite creatively, incorporating coloured words, images and a mechanism for readers to personalise the story by inputting their own characters—though their stories tended to be quite linear. The older students generally had stories with more complex storylines and multiple decision pathways, though they experimented much less with the styles and additional HTML/CSS/JavaScript functionalities.

Themes Emerging from Students' Stories

Informal analysis of the students' submitted IDNs revealed some recurring themes: tourism, related to the CO_2 impact for various modes of travel; plastic waste and pollution; and food choices, with an emphasis on veganism and low-meat diets. While transport methods and food choices featured in *No*

World 4 Tomorrow, it is also likely that some of these themes emerged from external national and global conversations on these topics, such as the plastics debate inspired by British Broadcasting Corporation's (BBC) *Blue Planet II* (2017), which highlighted the devastating impact of plastic pollution on marine ecosystems or the growing popularity of veganism as illustrated by 'Veganuary', wherein a charity inspires people to try a vegan diet for January (2020). Collectively, our informal analysis of the content of the IDNs suggested that the students had reflected on the general themes from Workshops 1 and 2, but that they had also drawn extensively on their broader understanding of environmental issues.

Our formal qualitative analysis revealed deeper themes related to student culture and social structures. Stage 1 grounded theory coding of the student-submitted IDNs identified two core categories of emergent themes: *flight* and *fight* reactions to climate change, such as taking action to combat the crisis or abandoning the planet. Stage 2 coding further refined these categories into six types of story-specific responses to climate change, which Bourdieusien analysis typed into either 'dominant' or 'bridging' themes:

Dominant:

- *Fight-Denial*: deny climate change, often demonstrating a lack of understanding;

- *Fight-Individual*: modify personal choices and behaviours;

- *Fight-Holistic*: modify personal choices *and* urge governmental action.

Bridging:

- *Fight-State*: urge government action without modifying personal choices;

- *Flight-Social*: leave Earth through individual effort but as part of a group;

- *Flight-State*: leave Earth through a government programme.

Dominant themes reveal aspects of the learner's habitus, demonstrating that students' own value systems were not in tension at that point, and revealing the type of agency (individual or structural) they felt possible or feasible. Bridging themes highlight areas where dialectical confrontation (Ingram, 2011) between the learner's values and the You and CO$_2$ programme values occurred, indicating points where the learner's understanding of climate change was evolving, and thus their engagement in the CCE programme deepened. While less common than the dominant responses, bridging themes illuminate areas where the learners have allowed for modification of their habitus, embodying and enacting the value system of the unfamiliar habitus, advancing their understanding that climate change is a social—and indeed, global—challenge.

DISCUSSION

The You and CO_2 programme adopts a multidisciplinary approach to CCE. By doing so, we have been able to break out of siloed teaching and allow students the opportunity to think more broadly about the social and moral context of climate change and its consequences. Through this STEAM programme, the science of climate change has been taught in a way that is relevant and societally engaged, bridging the gap between the classroom and the wider world. In addition, the programme has allowed students to apply knowledge of numeracy, English language, storytelling, computer programming and chemistry to a pressing global problem that will have a profound impact on their lives. We reflect here on how successfully the You and CO_2 programme has met the recommendations laid out in the introduction section of this chapter, discuss barriers to student engagement with the programme identified through qualitative analysis and outline positive steps forward for You and CO_2 in the future.

Comparison to Bicycle Model of CCE

As noted above, Cantell, et al.'s 2019 Bicycle Model of CCE incorporates four key recommendations for developing effective programmes. These encompass enabling the students to believe change is possible, to see humans as dominant contributors and, thus, responsible for resolutions; engendering hope for the future; and incorporating interdisciplinary techniques to help students both understand the underlying science of climate change as well as the complex socio-political contexts in which it occurs. This section evaluates the pilot phase of You and CO_2 for how it fulfilled these recommendations.

Emphasise That Humans can Change Society; Engage Students in Joint Positive Action

Workshop 1 activities calculating carbon footprints and re-imagining futures based on positive changes to these habits present students with the idea that humans can change society and that 'local, tangible and actionable' endeavours (Jorgenson et al., 2019, p. 165) including individual actions can lead to greater global change. Workshop 2's *NW4T* IDN incorporates active choices leading to various avenues of action, from individual to societal (as well as no action at all); this reinforces students' ability to change society either singly or through joint positive action. In Workshop 3, this lesson is extended through their own IDNs; whether or not students are working on a single IDN in a group, they nonetheless create them in a classroom environment, sharing their stories with one another and discussing their creations.

Encourage Students to Think about the Role of Human Beings as Consumers, and Therefore, the Cause of Environmental Problems

Workshop 1's carbon footprint activity drives this point home quite sharply; this is reflected in the stories they create in Workshop 3, where they often question their own consumer choices (such as the environmental effects of wearing make-up or manufacturing a soccer ball). These stories also demonstrate achievement of a key learning outcome: for students to develop critical thinking regarding climate change. They show they are able to take the basic concept of carbon footprint from their breakfast and school commute routines, and extrapolate it to other areas of their life, such as shopping, recreation and health.

Stimulate Hope and Compassion

Workshop 3 gives students the opportunity to write their own IDNs and these provide the strongest evidence that the You and CO$_2$ programme stimulates hope and compassion (Cantell et al., 2019; Jorgenson et al., 2019). Some students wrote about being leaders of cities/countries/the European Union and having decision-making powers that would mitigate climate change; others wrote about social justice, ecosystem effects of the cosmetics industry and how climate change will affect poor versus wealthy countries. You and CO$_2$'s approach—advancing awareness of climate change then encouraging open-ended thinking about how to address the problem—incorporates hope and compassion at a foundational level. Interactive narrative places readers in an interactive environment, encouraging them to think about their choices; in *creating* their own IDNs, students place themselves in positions of power in order to explore facets of the crisis and propose their own solutions. This in and of itself stimulates hope.

Combine Science-based Teaching with Critical Thinking, So That Students Can Assess Technological Advances within a Broader Social Context

Workshop 1 encourages students to develop science-process skills (Ardoin et al., 2018) regarding the carbon cycle and its underlying chemistry. Students are encouraged throughout all three workshops to research other causes contributing to and effects resulting from climate change. Workshop 2's IDN creates global society on a microscale (a small settlement on the moon), complete with political factions, technological issues and personal concerns, so that students can grasp the very large, globally complex issue of climate change through a smaller metaphorical lens.

Socio-cultural Barriers to Student Engagement with CCE

The six themes defined from qualitative analysis of the students' IDNs indicate varying levels of engagement in the topic; by recognising these themes, climate change educators can frame their programmes to mitigate student resistance originating in cultural differences. As defined in the Results section, these themes incorporate students' reactions to climate change (*fight* or *flight*); Bourdieusien-relevant 'dominant' or 'bridging' themes; and levels of action to mitigate climate change from the individual level to governmental. This section discusses how these findings expose socio-cultural barriers to the students' engagement in the You and CO_2 programme, and how we can address them.

Fight-Denial (Dominant)

Fight-Denial (Dominant) reflects a very shallow level of engagement with the topic; this theme reveals an underlying rejection of the relevance of climate change to the student's habitus. As climate change as 'urgent' and needing action is linked to middle-class values and habitus (Drewes, Henderson, & Mouza, 2018; Laidley, 2013), and students who do not embody these values experience a level of discomfort when confronted with them, leaving them unable to engage with CCE. They typically convey this discomfort by mocking the topic, emphasising this habitus clash by acting contrary to social rules of their setting (school) (Ross, Rudd, Skains & Horry, 2021). This mockery is demonstrated in their IDNs via the tone of their language used or the absurdity with which they fulfilled the task.

School 2 responses were more evenly distributed between the dominant themes, with the largest proportion of responses lying in the Fight-Denial category. The high number of Fight-Denial responses at School 2 indicate that either the students did not understand the material or found it inaccessible (Ross et al., 2021)—the submitted IDNs displayed few faults, suggesting the topic area itself presented a barrier, rather than Twine as a new educational tool. Alternatively, the programme design did not sufficiently incorporate the cultural habitus and values (Pruneau, Khattabi, & Demers, 2010) of the school's diverse student body. While students at the school generally come from very affluent backgrounds, 32% are international students and, thus, their familial or cultural values may not align with those of the school; climate change may not be perceived as an urgent problem (Frantz & Mayer, 2009). The dialectical confrontations between their existing habitus and values (Bourdieu, 1984) and those encouraged in the You and CO$_2$ workshops were likely too significant, causing distress and lowered engagement in the students. This result reflects Laidley's 2013 argument that CCE undertaken without consideration of cultural and social values is ineffective.

Flight-Social (Bridging)

In IDNs with Flight-Social themes, protagonists' individual agency drove them to flee with a group to a safe location unaffected by climate change, demonstrating student exposure to a habitus they cannot engage with. These IDNs were most commonly group-authored, suggesting that the formation of a subgroup allowed students to collectively embody a habitus promoting personal action to combat climate change. Positioning themselves (as reflected through their protagonists) gave the students value (McKenzie, 2015), bridging their lack of understanding of the topic with an investment in the group habitus. The space for the modified habitus was situated within the subgroup rather than within the individual; thus Flight-Social themes provide a key bridging stage for climate change educators, indicating a barrier to understanding the topic even while students attempt to engage more deeply. By addressing this barrier, CCE programmes can enable students to progress towards a more active engagement with climate change.

Flight-State (Bridging)

Although Flight-State presented in only one student IDN, it demonstrates an important transition between the types of agency students perceive. In this particular IDN, the students demonstrate some understanding of climate change and make the step of drawing on governmental action as a means to combat climate change. For example:

> *In six months the government will extract all the air from the atmosphere, move all sources of food and drain all sources of water and take it to the shuttle, to send of 1/3 of the population to Saturn...*

The action proposed here suggests that these students view climate change as urgent (Frantz & Mayer, 2009) but do not perceive individual action as possible. The value systems and power structures of educational institutions are largely white and middle-class; likewise, CCE programmes' habitus holds that individual or collective action can mitigate climate change. Students whose value systems and habitus do not coincide with their educational institutions are less likely to embody the associated CCE-related habitus; thus, their response may instead defer to governmental structures to act to combat climate change. The Flight-State theme offers a bridging opportunity for CCE programmes to address this conflict and enable diverse learners to adjust their habitus to encompass more individual agency.

Fight-Individual (Dominant)

Fight-Individual IDNs show active engagement with CCE generally and the You and CO_2 project specifically. Thematically, this theme comprised the highest proportion of IDNs at School 1, suggesting these students understand that climate change is an urgent problem needing resolution and believe their own actions can mitigate it. School 1's affluent and academically successful nature suggests that the values within the school community align with those of white, middle-class hegemonic groups (Bourdieu, 1984; Tomlinson, 2011); likewise, these groups and their habitus are commonly linked to CCE and action (Drewes et al., 2018; Laidley, 2013). Therefore, the school and its students likely share a habitus, resulting in less likelihood of a dialectical confrontation resulting from a habitus clash between student and CCE programme. Where familial and educational habitus aligned, students could embody the 'appropriate' habitus for their school setting (Bourdieu, 1977), engage with the You and CO_2 project topic matter and values, embody them and reproduce them in their IDNs.

Fight-State (Bridging)

Fight-State responses demonstrate a shift in student understanding of governmental structures' role in mitigating climate change. The student has bridged the gap between their own habitus and the CCE habitus, enabling them to embody the new habitus (Ingram, 2011) in which government engagement is a positive step. Like Flight-State, Fight-State only arose in one story. While the IDN's main focus of agency is governmental, the text also suggests a potential role for individuals and their actions in addressing climate change:

...i would like to invite my people to the castle to have a lovely dinner and talk about the issues of the country with them so try to get some opinion [sic].

Congruent with Laidley's (2013) work, Fight-State responses highlight student understanding of the importance of governmental action to combat climate change. This type of response also suggests that the student can reconcile clashes between their home and school habitus, so they can fully engage with CCE. The lack of individual agency shown in the IDN, however, indicates that the student does not embody a habitus where individual and governmental action act to counter climate change; they still have room to progress to full engagement with and embodiment of the You and CO$_2$ programme's habitus.

Fight-Holistic (Dominant)

Fight-Holistic IDNs demonstrate a good understanding of CCE, active engagement with the You and CO$_2$ project, and understanding that climate change can be countered through *both* governmental and individual action. Several School 1 IDNs expressed Fight-Holistic responses, showing awareness of the roles of both individuals and governments to mitigate climate change. This holistic approach suggests that their familial habitus aligned with the school habitus, which has expectations of engagement with wider institutions, access to decision-making processes and positive interactions with all members of the community (Estyn, 2018). Thus, the students embody a habitus that facilitates interaction with institutions.

Implications and Recommendations

Combining the coded themes identified in this study's qualitative analysis and Cantell et al.'s 2019 Bicycle Model for CCE presents a practical, tangible understanding of the different stages of CCE and the elements that facilitate student transitions from one stage to another. The stages represent milestones of pedagogical achievement that can be identified in student work, from Fight-Denial to Fight-Holistic. The pathway the learner travels between the stages represents the Bicycle Model's tools that enable educators to adjust the CCE programme elements to the particular needs of the student and/or student group. Incorporating the Bicycle Model offers CCE educators programme elements that are within their remit to change/modify, such as hope and emotions (the lamp), motivation and participation (the saddle) and operational barriers (the brakes). Together, these form the framework of our 'holistic Agentic Climate-Change Engagement' model (h-ACE), illustrated in Fig. 2.3.

The h-ACE model provides a contextualised, theoretical understanding of the identity, values and world view (frame) underpinning students' experiences and value systems. Awareness of these values and frameworks enables CCE educators to engage with their students' habitus. In turn, this heightened awareness enables teachers to support their students through any dialectical confrontations (Ingram, 2011) that occur as a result of exposure to and potential embodiment of a new habitus regarding climate change. Utilising CCE programmes that incorporate the h-ACE model empowers students to transition to a habitus in which climate change action holds value (Drewes et al., 2018) and move towards a paradigm in which action is possible and desirable (Frantz & Mayer, 2009).

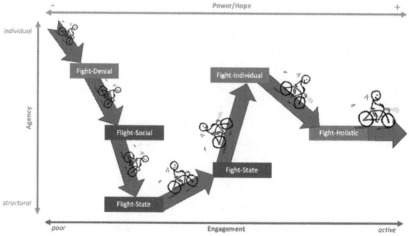

Fig. 2.3. Holistic Agentic Climate-Change Engagement Model (h-ACE).

Transitioning You and CO$_2$ from Trial to Holistic CCE Programme

The You and CO$_2$ project aims to empower young people to act to combat climate change, both at individual and governmental levels. In this, its first iteration, we based its design primarily on the New Welsh Curriculum (Welsh Government, 2015), using the three workshops and student reading and writing of IDNs to encourage students to: (1) commit to a sustainable planet, (2) understand and exercise their individual and structural agency and (3) understand and consider the impact of their choices and actions. The initial design of the You and CO$_2$ programme embodies and propagates a value system where climate change is an urgent problem requiring action; this is an inherently middle-class value (Drewes et al., 2018; Laidley, 2013; cf. Bourdieu, 1984; Tomlinson, 2011; Welsh Government, 2015) which may conflict with the values of those who are not white and middle-class.

The student-submitted IDNs reflected the students' educational contexts. School 1 students were positively engaged with CCE and had a positive view of individuals' ability to mitigate climate change. This reflects the school culture as a high-performing school, embodying middle-class habitus and value system. A large proportion of responses also engaged at government level, which mirrors the school ethos, where students' access to 'power' within the school is encouraged. Where students' values did not align with white, middle-class values, a larger proportion of their IDNs displayed denial and disengagement. This aligns with Laidley's (2013) view that engagement and action on climate change is significantly affected by social class and culture; this appeared to be the case at School 2. Given the high number of international students at School 2, however, it is likely these students' dialectical confrontation was not that of a clash between middle- and working-class values, but rather between white, UK-based, middle-class values and those of the international students' home nations. As such, further work is needed to explore the responses of working-class students to CCE programmes such as the You and CO$_2$ project. That said, the h-ACE model can help teachers and educators understand how to plan lessons or programmes of study so that such cultural- and values-based barriers can be minimised.

As noted in the Methodology section, the You and CO$_2$ programme delivery differed significantly between Schools 1 and 2 in terms of length of the workshops, time between workshop deliveries and pedagogical approaches to Workshop 1. These differences in programme delivery mean that any conclusions drawn from their data must be clarified through further work, in different settings but with consistent delivery. Similarly, the high proportion of incomplete IDNs indicates a structural inadequacy with the first iteration of the programme: insufficient time. Students were not given enough time to complete their IDNs; likewise, for those with additional learning needs, the material is not yet adequately differentiated to meet their needs. Further work with teachers and other professionals is necessary to improve this for future delivery of the programme.

It is also important to note that this initial iteration of You and CO_2 did not collect demographic data from the students, so it is not possible to make robust, secure claims regarding the individual students' familial habitus and how these are reflected in their IDN responses. Future instances of the You and CO_2 CCE programme must collect these demographic data and map them against individual IDNs in order to better contextualise the student responses. Likewise, implementing You and CO_2 in a spectrum of schools with varying demographics in terms of economic class and cultural background will offer deeper insights into the role of habitus with CCE engagement. Each of the schools in this study had a significant proportion of students from white, affluent backgrounds, which does not give deep insight into the experiences of other demographics.

Revisions to the You and CO_2 CCE programme will, thus, mitigate these limitations. Changes will be incorporated to increase programme accessibility for students with varying educational needs. These may include additional optional 'modules' for teachers to implement on an as-needed basis, such as extra time within and between workshops, IDN planning exercises and templates for constructing their IDNs. The workshop design will be modified, so that the introduction and promotion of its CCE value system does not act to constrain and actively oppress those whose value system and habitus does not or cannot incorporate the 'new' You and CO_2 habitus. Failing to reconcile these potential dialectic confrontations risks alienating those students whose values do not align with the 'accepted' middle-class ones, leading to the formation of 'subgroups' who refuse to engage with climate change at all. This may include localisation of the curriculum, to better reflect the students' local environment, familial habitus and knowledge structures. For increased robustness of the data set, demographic data will be collected in order to draw a clearer picture of individuals' specific habitus and how these are reflected in their IDNs.

CONCLUSIONS

You and CO_2 is a STEAM programme combining chemistry and CCE with IDN-based bibliotherapy and expressive writing to engage young people with the social and moral context of climate change and to encourage them to consider their own role as a consumer and as a member of a wider society. Our multi-disciplinary approaches allow the students the chance to explore themes around climate change in a novel and engaging way. This is reflected in the IDNs written by the students, which have diverse storylines but featured common themes for reducing one's carbon footprint, including travel, food consumption and plastic waste.

The combination of all three You and CO$_2$ workshops introduces the science of climate change on a personal level, then encourages students to approach the issue from a personal, actionable perspective. Done in a classroom setting, these activities encourage group interaction and feedback, and students often work together to design solutions to climate-related aspects of their own lives. Likewise, the open, creative nature of designing interactive IDNs in Workshop 3 encourages students to seek out answers to their scientific *how* and *why* questions, performing research to enhance their creative work. Overall, the You and CO$_2$ project pilot offers a cohesive, integrated STEAM approach that meets Cantell et al.'s recommendations as outlined in their Bicycle Model of CCE.

The You and CO$_2$ CCE programme is already a unique and engaging approach to CCE, demonstrated by the positive responses to it from educators and students, and the various institutions that have sought to incorporate it in their curricula. The study outlined here illustrates its current limitations and offers a clear and concrete path to improve the programme for deeper engagement and more thorough efficacy. You and CO$_2$'s foundation is strong, based as it is on the New Curriculum for Wales; with the developments outlined here, it can become an even more positive and powerful model to engage young people in the fight to battle catastrophic climate change.

3

PILOT CASE STUDY: INFECTIOUS STORYTELLING

INTRODUCTION

The Infectious Storytelling project, like You and CO_2, arose from Welsh Crucible collaborative exercises.[1] In this project, four rather disparate disciplines came together to address a 'big question' from a unique and innovative perspective: how can insights in arts, humanities and creative practice contribute to resolving a global health issue? The inspiration came from Dr Emma Hayhurst, our project microbiologist, and her passion about antimicrobial resistance (AMR). AMR is a problem that biologists and medical researchers are investigating worldwide, in terms of seeking new pharmaceuticals and treatment techniques. Yet, addressing one of the critical contributing factors to AMR—that of human behaviour—has mainly been left to health organisations like the World Health Organization (WHO) to 'campaign' and 'raise awareness'. Those of us who were interested in AMR as a topic of exploration brought our various perspectives: micro-biology, film and television media, Romantic art and literature, and digital creativity. From these, we surmised we could approach questions about AMR from a more creative media-based angle, examining how infectious disease has been treated in popular media of relevant periods, and how these portrayals affected people's behaviours related to the disease and its treatments. Given our fourth area of expertise—digital creativity —we then designed this project, Infectious Storytelling, wherein we could apply those lessons learned to a new work of narrative art, one whose purpose was to alter the public's behaviours that contribute to the increasingly urgent problem of AMR.

---

1 A pan-Wales research enterprise that connects and trains researchers across Wales in all disciplines. http://welshcrucible.org.uk/.

The WHO identifies AMR as one of the biggest challenges in global health; without immediate and coordinated action, we are facing a 'post-antibiotic' era, in which common infections could once again kill (2015). One of the WHO's top priority objectives is to 'improve awareness and understanding of AMR through effective communication, education and training' (2015, n.p.). It found, however, that awareness and understanding do not always lead to behaviour change (2017, p. 5); key groups related to antimicrobial use (which they identify as patients/public, prescribers, pharmacists and policymakers) continue to engage in behaviour such as overprescribing and pressuring doctors for unnecessary medication, despite lessons to the contrary.

The Infectious Storytelling project in the long-term anticipates developing educational measures for various aspects of the four 'P' groups concerned with AMR: public, prescribers, pharmacists, policymakers. In this first pilot phase, we have focused on effecting positive behaviour change in the first 'P': the public (patients). Inappropriate antibiotic prescribing is well known to be exacerbated by public pressure (Ashworth, White, Jongsma, Schofield, & Armstrong, 2016). Campaigns are currently underway to alter patient expectations and pressure, most notably the WHO-led World Antibiotic Awareness Weeks (WHO, 2019) starting in 2016 and the Antibiotic Guardian campaign launched by Public Health England in 2014. To date, however, little data are available as to the efficacy of these campaigns with the public (cf. Ashiru-Oredope & Hopkins, 2015; Bhattacharya, Hopkins, Sallis, Budd, & Ashiru-Oredope, 2017; Brinsley, Sinkowitz-Cochran, Cardo and the CDC Campaign to Prevent Antimicrobial Resistance Team, 2005; Newitt et al., 2018); the Antibiotic Guardian campaign primarily engaged with health-care practitioners and others likely to have pre-existing knowledge of AMR (Chaintarli et al., 2016; Kesten, Bhattacharya, Ashiru-Oredope, Gobin, & Audrey, 2017).

The goal of the Infectious Storytelling project is not to compete with or replace these campaigns, but rather to complement them by exploring another mechanism to raise awareness and firmly embed the desired behaviours in the target populations: interactive storytelling, specifically, interactive digital narratives (IDNs). IDNs have been shown to increase the efficacy of teaching on range of topics (Ensslin et al., 2016; Ernst & Colthorpe, 2007; Izzo, Langford, & Vitell, 2006), as they incorporate a multiliteracies pedagogical approach (Cope & Kalantzis, 2009a; Skains, 2019a) embedding the concepts on multiple cognitive levels. Likewise, popular media such as television shows, movies and games have been shown to affect audience behaviour (Bouman, 2016; Glik et al., 1998; Goldstein, 2015; Institute of Medicine, 2015; Singhal, 2013; Valente & Bharath, 1999; Whittier, Kennedy, St. Lawrence, Seeley, & Beck, 2005), particularly when carefully researched and aimed at target audiences.

Thus, we hypothesised that an IDN designed for both entertainment and education could be an effective tool to raise awareness of AMR leading to positive behaviour change; the project described here is the pilot phase of that effort. The key aims of the Infectious Storytelling project (in its pilot phase) were:

- To review the medical history of tuberculosis (TB) as a human disease, from its first origins through to the present-day issues of multidrug resistance.

- To examine representations of TB in British literary and visual culture since 1790.

- To examine socio-cultural responses to these representations regarding patient behaviour.

- To draw upon these conclusions to create an IDN for patient education.

- To evaluate the feasibility of the IDN as a method of patient education.

As this chapter reports only on the pilot phase of the project, we do not yet have data regarding the long-term effects and efficacy of the programme; we reserve our speculations regarding such for later work, once we are able to apply the lessons learned from the pilot phase and expand the programme to a much larger scale in terms of participants and time. The following sections describe the activities we undertook as part of this pilot: our initial research and analysis methods, the results of such, and our conclusions based on these results prior to adjusting the programme for the next phase of expansion.

METHODS

'Infectious disease' and 'AMR' are both incredibly broad terms covering a wide array of pathogens, including but not limited to bacterial and viral infections. We narrowed down our case study disease to one that was both applicable to all of our research areas and that was of contemporary concern: TB. TB is ideally suited to this project because it has a long history and consistent presence in narrative media and is a current AMR health concern; multidrug-resistant tuberculosis (MDR-TB) is a disease affecting ~600,000 new people yearly (WHO, 2018). As patient pressure leading to overprescribing primarily revolves around antibiotics (as opposed to, for example, antivirals), a bacterial pathogen case study was preferable; likewise, its history as a disease with an antimicrobial cure and subsequent resurgence as a resistant strain suits the project's message well.

We obtained ethical approval for this project from each of the participating universities: Aberystwyth University, Bangor University, Cardiff Metropolitan University, University of South Wales and, later, Manchester Metropolitan University (as the Primary Investigator (PI) had transferred post). Using representations of TB in literary and visual cultures since the 1790s, our background research investigated how these representations have altered public attitudes towards the disease and its treatment. Our specialist in Romantic literature and art, Dr Carmen Casaliggi, reviewed these representations and the current discourse on TB's role in art and literature of the long nineteenth century (1789–1914), including how the public generally responded to them. Similarly, our film scholar, Dr Kate Woodward, reviewed films in which TB affected the story line or a significant character, and our digital/contemporary media scholar reviewed the more recent appearances of TB in fiction and games.

We also extended our research to a review of the National Library of Wales' archive of materials related to the Welsh National Memorial Association (WNMA) (Wales' anti-TB programme, 1911–1948). While such research may not have been strictly necessary, it was nonetheless useful, as it brought us much closer to the very human concerns of the pandemic, its victims, its heroes, their searches for treatments and cures, and their frustrations with the local and national governmental infrastructures related to these. These archives included articles, blueprints, medical reports, budgets, invoices and personal correspondence related to the WNMA.

We incorporated into these background studies our analysis of educational campaigns regarding TB and AMR, focussing primarily on campaigns prominent in the United Kingdom. The most prominent are Public Health England's 'Antibiotic Guardians' (2014) and the WHO's World Antibiotic Awareness Weeks since 2016, with the messages renewing each year (2019). We also examined public health campaigns unrelated to AMR yet that showed efficacy, including television drama campaigns for sexual health and sexually transmitted disease information, breast cancer awareness and malaria prevention (Fedunkiw, 2003; Glik et al., 1998; Hether, Huang, Beck, Murphy, & Valente, 2008; Valente & Bharath, 1999; Wang & Singhal, 2016; Whittier et al., 2005).

This research served as a foundation to the practice-based phase of the pilot: construction of the purpose-built IDN for educating patients about AMR. Our digital creativity specialist, Dr Lyle Skains, constructed a set of requirements for the IDN based on analysis of health campaigns, as well as analysis of the target audience (adult patients or caregivers making medical decisions for themselves and/or others), and their reading/playing context (general practice [GP] and/or hospital waiting areas) and devices (mobile phones using free Wi-Fi or mobile data). This set of requirements included:

- Short playthrough time (5–10 minutes), as appropriate for appointment waiting times;

- Use of fairy tale theme for quick connection to story world and characters (cf. Singhal, 2013; Sood, Menard, & Witte, 2003);

- Cross-platform mobile compatible;

- Light use of data-heavy items such as images, videos or animations (which delay download times while on mobile data or free public Wi-Fi);

- Repetition of key phrase/message as a mnemonic.

Once drafted, the IDN was beta-tested among the project investigators, and close friends and family for playability, cross-platform compatibility and errors.

The revised IDN was then deployed in focus groups. While the original proposal had called for these focus groups to be conducted in person at each of the investigating sites (Bangor, Aberystwyth, Cardiff Met and South Wales), the completion of the IDN coincided with the UK COVID-19 lockdown; ethics approval for online focus groups was obtained from the PI's new institution (Manchester Met). The investigators shared a call for participants on email and social media. Each volunteer received a Participant Information Sheet and gave electronic consent to participate. Focus groups were conducted and recorded on video conferencing software including Skype and Zoom.[2] At the start of the focus group session, participants filled in a pre-game questionnaire that primarily collected demographic data, along with one question to evaluate their level of familiarity with the appropriate use of antibiotics: 'Which of these conditions can be treated with antibiotics?' Options ranged from 'a cold', 'athlete's foot' to 'stomach ulcers' and 'COVID-19'.

Participants were then given a quick response (QR) code and uniform resource locator (URL) to the project IDN and asked to use their mobile phones to bring up and play the IDN in one and only one playthrough. Each playthrough of the IDN was recorded by participant number as links clicked. The researcher then led participants through a discussion of the game, focussing on its message, memorability and any barriers to accepting that message including technical or aesthetic issues. Participants then returned to the online survey

---

2 The researchers acknowledge the known security and privacy issues with both Skype and Zoom products. Attempts to conduct focus groups on more secure platforms, such as Skype for Business and Microsoft Teams, however, encountered immovable obstacles (namely, that either the researchers or the participants—or both—could not access these from their various institutions or locations). As no personal data were being shared on the video conferencing itself, we deemed the use of these less secure systems acceptable.

and completed the post-game questionnaire. This section focuses on questions about the participants' medical decision-making, as well as their professional roles that might include use of antibiotics (such as medical dispensing or working with livestock). The final question was a repeat of the antibiotic-appropriateness question from the pre-game questionnaire, as an immediate indicator of whether their knowledge and awareness of appropriate treatment for common ailments had improved as a result of playing the IDN.

Data analysis at this stage is limited to researcher observation and informal qualitative impressions of feedback during the focus groups, as well as initial analysis of the playthrough and survey data. Further analysis is planned for the second phase of the project, when funding can be obtained to transcribe the focus group discussions, and targeted analysis of anonymised transcripts, playthrough records and survey responses can be conducted.

RESULTS AND DISCUSSION

A Brief History of Tuberculosis

Our initial background research into representations of TB in popular creative media (poetry, literature, film and games) brought several interesting features to light. Arno Karlen (1996) offers a look into the path of TB as a human disease. The bovine infectious agent, *Mycobacterium bovis*, mutated to infect humans around 7,000 years ago, primarily affecting the bowel, lymph glands and spine (as expressed through the Pott's disease in the titular character of Victor Hugo's *The Hunchback of Notre-Dame*, 1834). The more familiar pulmonary strain, *Mycobacterium tuberculosis*, mutated from *M. bovis* ca. 4,000 years ago, spreading from East Asia to Greece and eventually the rest of Europe. As its cousin leprosy (*M. leprae*) rose in the Dark Ages (476–1453AD), it conveyed some level of immunity against TB, which declined until the Middle Ages. Increasing urbanization, and, thus, crowding and poverty set TB on the rise again, until by the eighteenth century, it was so widespread it became not only common, but strangely de rigueur.

Robert Koch identified *M. tuberculosis* as the TB disease vector in 1882 (Inglis, 1965). Less than 10 years later, he announced that he had a vaccine, which prompted a global rush for the 'tuberculin'; the craze was short-lived, however, as pathologist Rudolf Virchow proved the vaccine entirely ineffective (Reid, 1975). An effective treatment was not discovered until 1947, with the advent of antibiotics and specifically streptomycin (Karlen, 1996). The age of scientific medicine had arrived, and the popular image of TB soured, as sufferers were placed in sanitoria, subjected to cures from tonics to orchestrated lung collapse and restricted from specific occupations and locations (Morens, 2002, p. 1356). Interestingly, our home nation of Wales developed the only national institution for TB management, the WNMA (1911–1948); the WNMA archive at the National Library of Wales provides intimate insights into the structures, funding and attitudes related to TB treatment before antibiotic treatment was introduced, and it was absorbed into the newly developed National Health Service (NHS). For example, the developing technology of X-ray radiography was touted as a key diagnostic tool, and the WNMA recommended regular scans for all citizens; similarly, therapeutic pneumothorax (artificially collapsing the lungs) was touted as a positive treatment for TB (though it would later be shown to be ineffective) (Childerhose, 1936).

M. tuberculosis infections proved vulnerable to antibiotic treatment, which led to a significant decline of infections in the West by the 1970s. By the 1980s, however, TB re-emerged, providing us with an exemplar of antibiotic resistance: TB sufferers quickly feel better, within a few weeks, and adherence to the treatment regimen—taking meds—drops off. Incomplete treatment courses result in the most resistant bacteria remaining in the patient; these bacteria then replicate and reproduce their resistance. A few cycles of this, and the patient is now harbouring a colony of *M. tuberculosis* that cannot be eradicated using that medication. As *M. tuberculosis* is persistent and highly contagious, repetition of this process with various classes of antibiotics has resulted in its modern resurgence into a full-blown multi-drug resistant (MDR) pandemic: TB now infects nearly 2 billion globally; of these, 10 million have active TB, with 3 million TB-related deaths per year (WHO, 2018).

Popular Representations of TB

Given TB's long history as a human disease, it is unsurprising that it has been represented thousands of times in art, literature and media. In the long nineteenth century (ca. 1789–1914), the preponderance of TB-infected Romantic artists, including John Keats and (supposedly) Percy Shelley, created a myth that 'consumption' drove artistic genius. Like the more recent 'heroin chic', poets and artists of the age popularised a damaging and deadly disease. Lord Byron imagined his death by TB would appeal to women (Gordon, 1839, p. 113), and Alexandre Dumas noted that 'It was the fashion to suffer from the lungs… and to die before reaching the age of thirty' (in Dubos & Dubos, 1987, p. 58). The white complexion and drawn features became the reigning look, even for those lucky enough to avoid infection by a disease on the rise due to expanding urbanization, urban crowding and insufficient sanitation. Poets, artists and musicians centred their themes on this 'typical' Romantic disease, building the myth of the consumptive genius. For example, Shelley's *Adonais*, in memorializing Keats, made explicit the notion that inner creative passion or genius somehow burned too brightly for the fleshy casing of the body: 'the bloom... Died on the promise of the fruit' (Dubos & Dubos, 1987; Shelley, 1891). This *spes phthisica*, a kind of elation that intermingles with depression during the disease, was believed to elevate the mind and, thus, the art (Dubos & Dubos, 1987).

As science and the study of medicine grew to prominence in the latter half of the nineteenth century, however, public perception of TB changed. Koch identified the cause of TB not as a creative soul that burned too brightly for its mortal coil, but rather as a 'germ' that invaded and infected. Thus, TB became a plague to be defeated, rather than an affliction to be desired, particularly as more realistic portrayals of TB emerged in later examples of nineteenth century art and fiction, such as Tolstoy's *Anna Karenina* (published in 1878, but first translated into English and published in the United Kingdom in 1889).

The mid-twentieth century presents another key era for TB. As TB treatment shifted to the newly developed medicines of antibiotics post-1940 (Daniel, 2000), Hollywood film techniques and the films themselves were used in public health programmes and mobilizations regarding infectious disease (NIH, 2011; Institute of Medicine, 2015). Despite this WHO-identified global health challenge status (2018), TB's treatment in contemporary creative media as an historical disease; if it is acknowledged as a modern infection, it is largely presented as curable. Films featuring characters with TB are period pieces, placing the wasting disease firmly in the mythic history of the American West (*Tombstone*, 1993), the Romantic era (*Moulin Rouge*, 2001) or of scientific discoveries and medical cures (*The Citadel*, 1938). Medical dramas (*ER*, 1994; *New Amsterdam*, 2018) occasionally include a TB patient in episodes, usually immigrants or travellers who have received insufficient health care. TB has almost no expression in games and interactive stories, save as yet another historical disease in *Red Dead Redemption II* (set in the American West) (2018). While underlying fears of apocalypse by disease are frequently represented in films and games, whether through zombies or more fact-inspired outbreak tales (usually based on haemorrhagic fevers like Ebola or viral vectors like influenza: Stephen King's *The Stand* [1990] or 1995's *Outbreak*), the *actual* pandemic of TB is rarely represented.

Construction of Only Always Never

Review of successful health campaigns in fictional entertainment, as well as the contextual considerations for the work, significantly influenced the form and content of the IDN. Arvind Singhal et al.'s collection *Entertainment-Education and Social Change* (2003), in profiling several successful entertainment-education campaigns, offered key structures for success in health communication. These included the need to invoke intellectual, emotional and physical responses in the target audience; the audience must understand and accept the message, connect in some personal way to the characters or the message, and have a concrete action to carry out. The actions need to align with the audience's culture; a community will not embrace change that challenges the core of their world-view. Likewise, the changes the communicator is asking the audience to make need to be manageable; small, iterative changes are more palatable and real-istic to implement.

Most of the campaigns Singhal et al. profiled were ongoing television series (such as soap operas or dramas); the communicators in these cases had several factors in their favour built into the medium. The target audience was already invested emotionally in the characters, and the series were an integrated compo-nent of their community and culture. For the Infectious Storytelling project, we would not be able to benefit from those built-in components, as we intended to create and implement a new, stand-alone work. Suruchi Sood et al. (2003, p. 129) presented a suitable strategy: 'Common legends and folktales occurring within specified cultures can be used to develop powerful, emotionally charged entertainment-education programs that succeed in promoting value, belief, and behavior change'. Particularly as the context of our intended work—GP and hospital waiting areas—significantly limited the length of the piece, integrating a known story and familiar characters enabled a shortcut to emotional engagement.

The first author (Skains) made specific choices for *Only Always Never* based on the above background research in combination with the situational parameters for implementation of the IDN. She embedded the story within the known story world and characters of L. Frank Baum's *Wizard of Oz* series (1900), whose various adaptations and continued presence in popular culture make it instantly recognizable and emotionally engaging across the world, despite its distinct American setting (later books in the series saw characters travel to other locales, including London and Australia, and the 1939 film is an unassailable classic). It also offers scope for expanded iterations of the message, as both Kansas and Oz are dominated by farmland and agriculture; antibiotic overuse in livestock is a contributing factor to AMR, and the project's future plans include design of IDN(s) specifically aimed at farmers, livestock handlers and veterinary practitioners. Its release into public domain also meant we were freely able to use, adapt and alter the story and imagery to suit our message without concern for royalties or copyright holders.

The other key story element is the memorability of the message. AMR can actually be a complex-scientific concept to express, involving genetics, microbial replication, mechanisms of cellular destruction and more. Worse, the messages from the twentieth century culture surrounding the 'miracle' cures of antibiotics still persist, as many patients expect that whatever ails them, a pill or injection can cure it (a message reinforced by neat happy ending Hollywood cures produced, manufactured and distributed with magical speed in films like *Outbreak* and series like *The Last Ship*). It is not necessary, however, that the target audience completely understand the science of AMR in order to carry out the desired actions. The IDN, thus, focused on embedding a mnemonic—or rather, further embedding a mnemonic, as 'only always never' was pulled from the WHO's 2016 antibiotic resistance campaign. Thus, the message echoes a wider campaign and offers a recipe to implement the desired action: to only use meds as prescribed, to always take the full prescription and to never use leftover antibiotics.

We designed the IDN to be implemented in GP and hospital waiting areas. These are ideal areas to reach out to the target audience (patients), in a slice of their lives when they are both thinking about their health and likely at loose ends for something to do (as they are waiting). We envision a poster in the waiting area that draws the audience in and appeals to them to while away their time waiting for their appointment by playing a little game; an added benefit is that the poster serves to remind and reinforce the message once the IDN has been played. NHS surgeries and waiting areas usually offer free Wi-Fi, and a QR code and/or quick URL can direct patient phones to the IDN.

Only Always Never was, thus, designed to be light in terms of data downloads; free Wi-Fi is often slow and laggy, and mobile data inconsistent. It incorporates minimal images, and no video or animation. It is audio-free, given that: (1) in busy waiting areas, audio would be difficult to hear and (2) playing audio in public spaces is inconsiderate. The reader/player can progress through one playthrough in only a few minutes and no more than ten. The IDN is designed specifically for mobile devices rather than tablets or computers, as these will not be available to patients in waiting rooms. Finally, the IDN was pitched at a general reading level, to make it accessible to readers of all educational backgrounds.

Focus Group Insights

As of the date of this draft, we have conducted a feasibility round of focus groups, centred on evaluating the readability and appeal of the IDN, and usefulness of the survey questions. From these early groups, a few themes have emerged. Participants offered some technical feedback, primarily concerned with understanding the IDN and reading it. Participants in more than one group agreed that the yellow colour of some links (those leading the player/character down the yellow brick road to Emerald City) is difficult to see on their devices. Some participants expressed confusion over the transition from actual world to entering Oz, and several struggled in the first few passages to grasp the concept of the *cycling* links. Cycling links do not lead to a new page, but rather change their text each time they are clicked; e.g., the player/character can click on '*ignore* the yellow brick road' to change it to '*follow* the yellow brick road' before using another link to advance the story. The cycling links do affect where the story will go, and on the first playthrough, this was not apparent to the participants, and thus their recorded choices may not actually reflect what their *preferred* choice would have been had they understood this function.

In terms of content, all participants thus far reported very much enjoying the story, without a 'sticking point' where they would be inclined to leave the story unfinished. They connected to the Oz themes and characters, and by and large, agreed that the most memorable element of the IDN is the 'only always never' refrain. Several participants expressed surprise at learning the depth of the AMR issue, particularly with regard to MDR-TB; this emotion may be likely to more firmly embed the message, leading to both action and sharing with their social circles, an element Singhal et al. (2003) note aids success in health campaigns.

Survey responses thus far indicate that most of our participants-to-date are highly educated (six of nine have postgraduate degrees) and white. To date, there is a 30/60% female male gender split, and a range of ages from the 18–24 bracket to the 51–55 bracket (with a median of 41–45). This indicates a potential source of bias in the results, as this level of education is not representative of the UK population as a whole, as only 38% of the population has achieved a postgraduate qualification (HESA, 2020). Similarly, an entirely white cohort is not reflective of the diversity of the UK population.

One survey question is repeated before and after playing the IDN: 'Which of the following conditions can be treated with antibiotics?' Participants can choose as many as they deem appropriate from: a cold, the flu, sinus infection, chest infection, athlete's foot, stomach ulcers, glandular fever, TB and COVID-19 (of these, only sinus and chest infections, stomach ulcers and TB are caused by bacterial infections and thus can be treated with antibiotics). 78% of participants correctly identified sinus and chest infections; fewer (56%) correctly identified both stomach ulcers and TB. The 'wrong' answers all received at least one or two ticks, indicating some level of confusion as to how viral and fungal infections can be treated.[3]

Post IDN, the responses remained largely the same, though the percentage correctly identifying TB increased to 78%. These early results indicate remaining confusion about antibiotics, which may be irrelevant to the success of the project intervention; after all, if patients are convinced to *only* take meds as prescribed, *always* take the full prescription, and *never* use leftover meds, then it does not actually matter if they understand the *why* behind these actions. Future iterations of the project will likely revisit the construction of this question to provide a more accurate measure of buy-in to the central campaign message.

Some participants responded in the negative when asked if they would be likely to play this IDN in a GP or hospital waiting area, noting that doing so was just unfamiliar enough that they simply would not be in the correct headspace or mood to play a game. This indicates that the design of the poster hanging in the waiting area is a key component of the project and requires significant care and attention to entice the target audience into playing the IDN. The next phase of the project will need to encompass this aspect as well.

CONCLUSIONS

The Infectious Storytelling project has completed its pilot phase; this phase was designed as a feasibility study, yet a significant amount of research and insights have already been generated across multiple disciplines. It is clear from the background research that popular entertainment—whether that is poetry in the long nineteenth century, film in the twentieth or games in the twenty-first—has significant effects on the public's behaviour regarding health and disease. With well-designed campaigns, science and health communicators can achieve measurable results through entertainment media in patient behaviour and outcomes.

---

3 While viral and fungal infections can be treated by *antimicrobial* medications and these are also prone to developing resistance to their treatments, this project at this stage is focused on antibiotics treating bacterial infections.

This pilot phase has consisted primarily of background research, contextual research, practice-based research and feasibility testing of a specific intervention. Its long-term purpose is to establish a mechanism for health communication that effects positive behaviour change in its target audience. The primary innovation in this project is the use of *interactive* narrative to achieve these results; we hypothesise that active participation in directing and shaping a narrative can more effectively embed the desired message with the target audience. This hypothesis is supported by research into the pedagogical strengths of games for education.

The next steps for this project include revision of the IDN to address the technical concerns raised by the focus group participants: colours of links, smoothing of the transition to Oz and clarity of the cycling link mechanism. The survey questions will be revisited to offer more insight into the impact of the IDN's educational message, and the participant pool will be expanded to include a more representative sample of UK adults. We will implement broader quantitative methods to quantify the effects of this intervention. With additional funding and support, we plan to implement the IDN and qualitative data collection in more diverse settings, including festivals and county fairs where we can more reasonably expect to engage farmers and livestock handlers, as well as medical settings with NHS patients. We also plan to develop a more controlled trial of the intervention, creating a non-interactive prose story to implement in a control group to compare to the interactive narrative, in order to tease out the efficacy of the interactivity and participation. A further key aspect of this expansion will be long-term measures of the IDN's message engagement; we collect emails from all participants in order to measure their recall and antibiotic-related behaviour at longer periods after their experience with the IDN.

AMR is a pandemic that was hundreds of years in the making. Infectious Storytelling is a small cog in a large wheel of public health communication, yet our efforts to specifically examine and quantify its efficacy offer an innovation on the current AMR-related health campaigns. As part of a larger network of health and science communication through interactive narrative projects, including the You and CO_2 project (Rudd, Horry, & Skains, 2019), it provides a wealth of insight into the role of playable fictions and participatory stories in future public communication campaigns.

4

ENTERTAINING TO EDUCATE: CREATIVE AND PEDAGOGICAL INSIGHTS

INTRODUCTION

'Creative writing' is a practice that is primarily conducted as a solo adventure; we often imagine the romantic notion of the author locked away in their study, scratching away with a quill or furiously pounding a keyboard, bringing their visions to the page—even if the end result is a liminal product for a team to create, such as a play or film script.[1] Narrative and/or game design, on the other hand, is known to be a collaborative process, wherein the technology, the user interface, the affordances and limitations of the code and the budget and the hardware, all affect elements of any story that can be conveyed. The narrative designer in this instance must adapt the narrative to new directions from team members and product goals. For a writer moving from the first practice to the second, the changes and their resulting challenges can be dramatic.[2] Likewise, shifting to designing texts for educational purposes from a creative practice that is purely for the creator's gratification and satisfaction presents a challenge in and of itself. It alters the creative practitioner's rhetorical purpose significantly.

This chapter explores the challenges the narrative designer faces when writing to educational 'spec', or purposefully creating a text (in this case, an

---

1 All of which, of course, is a myth created by popular media. The Romantics, in particular, did not work alone; Percy Shelley had constant feedback from no less than Mary Shelley and Lord Byron, and vice versa. Henry David Thoreau had his mother and sister to feed him and do his laundry while he 'sequestered' himself in his 'hut'. And we'll never truly appreciate the fullness of Jane Austen's editor's contributions to her work.

2 This chapter refers to 'writer', 'designer', 'creator' and 'practitioner'; for the purposes of this discussion, these terms are used interchangeably.

interactive digital narrative, or IDN) to influence its audience for an educational and/or social purpose. As such, it is necessarily focused on the first author's (Skains) perspective, as she (hereafter, 'I') was the creative practitioner on both the You and CO_2 (*No World 4 Tomorrow*) and Infectious Storytelling (*Only Always Never*) projects. My specific area of creative practice (digital writing and narrative design) is primarily what inspired these projects to explore using IDNs for educational purposes; using my skills for health and science communication was a new direction. And while there is a wealth of material out there on how to write entertaining stories, and a growing body of research on science communication, there is very little instructional pedagogy on writing and designing creative texts that educate as well as entertain. There is certainly very little on designing *interactive* texts for this purpose; my task was to discover a working method for creating IDNs for health and science communication, and to use that to create IDNs for each individual project.

I use my established practice-based research methodology (Skains, 2018), in particular my Practitioner Model of Creative Cognition (Fig. 4.1), throughout this chapter to discuss the effects of changing my rhetorical purpose for these projects. Practice-based research entails a creative experiment designed to

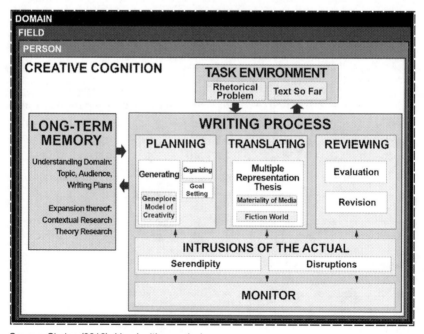

Source: Skains (2018). Used with permission.

Fig. 4.1. The Practitioner Model of Creative Cognition.

answer questions about the process and results of the practice itself: 'it involves the identification of research questions and problems, but the research methods, contexts and outputs then involve a significant focus on creative practice' (Sullivan, 2009, p. 48). The creative components of both You and CO_2 (YCO2) and Infectious Storytelling (IS) were conducted utilising this methodological approach, to answer questions about the creative process of designing IDNs for health and science communication. As such, their design process incorporated ethnomethodological (Garfinkel, 1967; cf. Brandt, 1992) observation of my writing activities, maintaining notes, journal entries, comments on drafts and other relevant, observable paratexts to the composition, in order to 'make continual sense… of what [the writer is] doing' (Brandt, 1992, p. 324). In this chapter I interpret these notes and paratexts, placing them within the context of my Practitioner Model of Creative Cognition, and incorporate media-specific analysis (Hayles, 2002) of the narratives that result. In this manner, the various strengths of practice-based research, ethnomethodology, cognitive process and textual analysis are combined into a robust method of evaluating my creative activities, and can be used to draw insights and recommendations for others looking to use IDNs for various communicative purposes.

As noted, the communicative purpose of both *No World 4 Tomorrow* (*NW4T*) and *Only Always Never* (*OAN*) was to effect positive behaviour change in the audience regarding the subjects of communication (climate change and antimicrobial resistance [AMR], respectively). On both projects, we examined existing educational materials on these topics, and determined that the primary goal for each narrative was that the audience connect emotionally and/or personally with the narrative. This initial impulse was borne out by subsequent communication research (Dahlstrom, 2014; Huffaker & Calvert, 2003; Rapeepisarn, Wong, Fung & Depickere, n.d.; Resnick, 2004; Singhal et al., 2003; Sood, Menard, & Witte, 2003). Our agreed approach was that the IDNs should be something their respective audiences would *enjoy* and seek out, regardless of in-class assignments or other types of required reading tasks. Enjoyable, entertaining narratives persist and spread via word of mouth, which similarly strengthens the communication potential of any embedded educational message.

The following section discusses the research contexts of the IDNs, and how awareness of previous entertainment-education projects shaped the process; this leads to a directed examination of writing to educational spec, with recommendations for future practitioners. The chapter rounds out with notes on the challenges of completing these projects, and insights into how these challenges can lead to more productive creative practices and effective IDN-based health and science communication.

DESIGNING ENTERTAINMENT FOR EDUCATION

The introduction to this book reviews the various approaches to using entertainment narratives for educational purposes that we have used as the foundation for both YCO2 and IS: bibliotherapy, expressive writing, narrative threads on mass media entertainment, and persuasive communication. This section offers a more specific account as to how I incorporated these aspects into the design of NW4T and OAN, focusing primarily on the latter two as these are most applicable to the *writing* of IDNs for health and science communication, rather than the teaching and implementation with target audiences.

Both projects are founded on the premises of bibliotherapy (Barker, 1995; Doll & Doll, 1997; Pardeck, 2014) and expressive writing (Pennebaker, 1997), as I had worked with these approaches on the Transformations project (Ensslin et al., 2016), which specifically used IDNs (termed in the project as *digital fiction*) for bibliotherapeutic and expressive writing interventions in teens with body image issues. On that project, I co-led a group of nine 16–19-year-olds in a two-week summer school focusing on body image issues, on which they each created a Twine game based on their own experiences of body image issues. The early stages of that project demonstrated a strong potential for IDNs/digital fiction in terms of therapeutic intervention, so when I became involved with collaborators who wanted to effect change in other topic areas through persuasive and/or educational communication, I funnelled those experiences into the design of both YCO2 and IS.

In terms of mass media entertainment, I am interested in how TV, film and games influence the public's knowledge in general; on the other hand, I don't actually have a role in the creation of any mass media entertainment. My rationale for taking this 'entertainment with education as a bonus by-product' approach instead of directed campaigns (of the anti-smoking, 'Just Say No' to drugs, or current COVID-safety variety) or 'edutainment' (of the teach-typing-games, *Sesame Street*, or afternoon special variety) is largely a factor of my own training and background. As a practitioner, I am skilled and experienced in writing narratives for entertainment purposes; I do not have expertise in developing educational materials, nor do I have access to mass media methods of distribution. The latter is particularly of note, as directed campaigns rely on the ubiquity and repeated repetition of their message. Without this ability to achieve a strong 'tipping point' for our message through pure media saturation, we would have to attempt a different approach. Given the creative practitioner involved in the project (me), the most appropriate route was to create entertainment media that *accurately* conveyed a science/health-based message. By creating a narrative with entertainment at its core, my aim was to influence readers to connect to the story on a deeper, more emotional and/or personal level, to counteract the lack of message saturation with a more profound connection to the narrative. This connection promotes continued and repeated engagement and sharing through word of mouth, significant factors in long-tail success of media properties (Anderson, 2006; Brown & Adler, 2008).

Finally, I drew on my training in persuasive communication as well as research on educating through entertainment media to incorporate effective communicative strategies at the core of both *NW4T* and *OAN*. Of particular use were Arvind Singhal et al.'s (2003) strategies for educating audiences through entertainment media, focusing on relatable characters, message saturation through repetition and word of mouth, iterative change, self-efficacy and compliance with cultural values and norms. Likewise, Suruchi Sood et al. (2003) specifically suggest the use of folktales in the narrative, as these stories are already embedded in the audience's culture, and their familiarity fosters emotional connection to the characters and the message. Folktales also enable the communicator to employ the highly effective 'given-new' rhetorical strategy, in which the delivery of new information is sugar-coated with familiar or 'given' knowledge; this technique enables familiarity and thus emotional and cultural connection, soft-touch iterative change and a feeling of self-efficacy for the target audience. This approach was very useful for *OAN*, as the time limits on that IDN left little room for new character development.

In the context of the Practitioner Model of Creative Cognition, this background work and research lies in the area of Long-Term Memory, expanding understanding of Csikszentmihalyi's *domain* (2006), or subject area: in this case, health and science communication, particularly through entertainment media. As noted, the rhetorical problem also shifted from the 'typical' creative purpose of creator satisfaction to specifically creating texts that would educate and persuade their audiences on given topics. Likewise, the collaborative nature of these projects ensured more intrusions of the actual, in an array of forms from team member expectations to serendipitous contextual research from others. These alterations introduced constraints with significant effects on the writing process, especially the planning and review phases, as discussed in the next section.

WRITING TO EDUCATIONAL SPEC

My creative process is a hybrid one as most writer/designers categorise them: I both plan my narratives and allow for some off-the-cuff creative exploration during drafting (cf. Skains, 2017); the 'Fiction World' box in the Practitioner Model accounts for this exploration during the translation phase. Planning generally focuses on characterisation and conflict that complements and highlights that characterisation; I don't typically concern myself with target audience needs, likes, reading levels or conveying specific messages with the work. Like many artists, my creative work is first and foremost for *me*; if others like it, that's great, but my initial goal is to create something I like, regardless of others' preferences or opinions. For YCO2 and IS, however, my rhetorical problem entailed writing persuasive, entertaining texts for specific audiences, in order to convey specific messages. This shift resulted in changes to my writing process in almost every aspect from planning to monitoring and review.

The following sections discuss the various impacts to the stages of my creative process, according to the Practitioner Model of Creative Cognition (cf. Flower & Hayes, 1981). While every element of the process experienced alterations, the two most significant were Planning and Intrusions of the Actual, due to the changes in the rhetorical problem (as discussed above) and the collaborative nature of these projects.

Long-term Memory

In addition to boosting my background knowledge in the areas of climate change, AMR and health and science communication more generally, I also researched the particular audience needs for each IDN, which ranged from reading levels to publishing trends to digital literacies and accessibility. As I don't usually write for a particular audience, much less a specifically Young Adult audience, I dug into popular texts across media for YCO2, identifying popular genres, tropes and language styles. These included popular texts (often transmedia) from the *Maze Runner* series (Dashner, 2013), *Simon vs. the Homo Sapiens Agenda* (Albertalli, 2015), the *Ship Breaker* series (Bacigalupi, 2010) and *Deadpool* (Nicieza & Liefeld, 1991). Key themes included humour, a slightly older protagonist (e.g., a 16–18-year-old protagonist for a 13–15-year-old target audience), localisation and personalisation (for personal connections), social media on various platforms, love interests (not necessarily heterosexual), self-realisation, feelings of control, and mystery/adventure.

The Infectious Storytelling project required less research on my part; the benefits of working in a team aided me a great deal here, as my project partners provided a wealth of research on tuberculosis (TB) and health campaigns from the Romantic period to present day (see Chapter 3). I supplemented their work with a paltry list of more contemporary references to TB (which continues to be treated as an 'old-timey' disease, and thus has little representation in current media beyond the game *Red Dead Redemption II* [Rockstar Studios, 2018], set in the American Old West, and minor one-episode storylines in various medical dramas). As the audience for the IS IDN was a more general adult audience, little specific audience research was necessary. It was the Planning phase of this project that truly required persistence and patience.

Planning

The planning phase in both YCO2 and IS was significantly affected, primarily in the box titled 'Geneplore Model of Creativity'. The Geneplore model (Finke, 1996; Ward, Smith, & Finke, 1999) encapsulates idea generation within the larger Practitioner model. It describes a cycle of three phases: (1) development of 'preinventive structures', or the germs of ideas; (2) exploration of these structures and (3) evaluation of these structures according to project constraints, leading to new preinventive structures and another cycle. This cycle repeats until, finally, an idea is generated that meets the goals of the project, falls within project constraints and motivates the creator to move forward into the Translation phase.

For the YCO2 project, as noted, I compiled a list of texts popular with the target age group (based on *Publishers Weekly* rankings and sales charts), and identified themes that resonated with this audience. I also noted themes that the project team had agreed upon based on our preliminary research into climate change education. Thus I had a list of constraints for the IDN I would create:

- Must be relatable to the audience (localised, personalised, in line with culture)

- Must allow the player-character to make simple choices that relate to everyday actions concerning climate change, which affect the outcome of the IDN

- Must incorporate a feeling of hope (so the player-character feels they have control and self-efficacy)

As a team we agreed that the idea of climate change is a very large one, a global problem; we hypothesized that this enormity led to a diffusion of responsibility (Wegner & Schaefer, 1978). If 7 billion people are responsible for climate change, how can any one individual's actions make a difference? Thus I wanted to create a 'microcosm' of our global issue, to help it relate more closely to single individuals and their actions. My preinventive structure, or idea germ, was to create a small, bounded society with strictly regulated resources, such as a totalitarian regime that segmented small bits of population into parcels of land, or a biodome of some sort. I wanted an artificial construct that would both limit resources and amplify the actions of individuals. Adding in a 'villainous' faction who hoarded or inappropriately used these limited resources gave the player-character actions regarding both their own responsible use of resources and their interactions with their society as to *society's* use of resources, mimicking both the personal choices regarding carbon footprints and the possibility of activism, voting and other societal pressures the audience can enact to bring about broader societal change. Given these constraints and developing pre-inventive structures, I fleshed out the final storyworld and narrative outline using a worldbuilding template I often employ when creating texts with alternative worlds, and transitioned to the Translation phase.

The Planning phase for the IS project was much more strenuous, cognitively speaking, despite (or perhaps *because of*) its broader audience with fewer specific needs. The team's background research identified three key audience segments for AMR education: prescribers (including doctors, pharmacists and veterinarians), policymakers and the public (including patients and agricultural workers). This first instance of the project is aimed at the last: primarily at patients, and those making medical decisions for others (such as parents for their children). Yet we hoped also to appeal to that second segment of the public, farmers, whose use of antibiotics in livestock contributes a great deal to AMR, through the food chain as well as run-off into the environment. Our serendipitous selection of TB as a case study aided us here, as it can infect both humans and livestock; attempting to incorporate both audiences, however, presented a challenge.

Likewise, IS carried specific constraints that had to be considered. Given the lack of mass media capabilities, we aimed to develop an IDN that could be delivered in an environment where the audience was already primed to accept health communication: GP surgery and hospital waiting areas. Reading/playing an IDN as a 'boredom buster' would be an ideal delivery system for a simple message about AMR-related behaviours; yet this ideal system placed constraints on the length, style and delivery capabilities for the IDN. In this reading context, one playthrough would need to be possible in the typical 5–15-minute waiting time (Ipsos MORI, 2020, p. 17), while permitting multiple playthroughs if waiting periods permitted. The IDN would need to be optimised for mobile devices, as we could not expect PCs or laptops in this context. It would also need to be easily accessible and downloadable across a range of mobile devices on either free public Wi-Fi or mobile data; these requirements called for a browser-based IDN with a very light data footprint.

These constraints meant that many of the techniques I had employed in *NW4T* were not possible: personalising the player-character's name and companions was too time-intensive and finicky for a reader-player on a mobile device with limited time. The short play-time also limited emotional connection, as 5–15 minutes is a difficult time period to establish character and storyworld enough for the audience to care about them; it also limited the complexity of the message, and the opportunity to embed a simpler message through repetition. A limited download size called for a largely text-based IDN, with little multimedia incorporated. Yet I had the same list of 'must haves' for this IDN as I did for *NW4T*: relatability, active choices, self-efficacy, and feelings of control and, of course, delivering an effective message about AMR-related behaviour to a general audience. This is a great deal to accomplish in a very short space of time and data.

The first piece of advice that enabled me to move forward in the Geneplore cycle came from Sood et al.: 'Common legends and folktales occurring within specified cultures can be used to develop powerful, emotionally charged entertainment-education programs that succeed in promoting value, belief, and behavior change' (2003, p. 129). This echoed something I had picked up at a talk from transmedia narrative creator Lance Weiler at Manchester Metropolitan University (2018), when I asked him why he so often adapted texts in the public domain; he replied that (1) they were free to use and (2) works in the public domain were already familiar to the public, and thus he could introduce new and innovative ways of telling stories in a more comfortable narrative setting (again, that 'given-new' technique at work) (Haviland & Clark, 1974). My creative gears churned and spat out a rather predictable answer, for anyone who knows me: L. Frank Baum's *The Wonderful Wizard of Oz* (1900), which despite its American origins is a beloved modern-ish folktale around the world. It has an easily recognisable storyworld and set of characters, and even incorporated farm settings for me to accommodate a secondary agricultural worker audience.

This was a serendipitous stroke, yet it still took several days and over eleven pages of brainstorming notes before I could land on a workable storyline and repeated motif that would serve as a mnemonic for the story and its message. The story used the storyworld and characters of Oz, but also incorporated research and knowledge about TB and its history in society, art and medicine. Drafting the mnemonic took me forward into the Translating phase, as I aimed to have that motif established prior to drafting the IDN, so that I could embed it over multiple levels as deeply as I could. I dug into previous

WHO and Public Health England campaigns, and took a 2016 campaign's repeated use of 'only', 'always' and 'never' to create a magic spell to incorporate throughout the story. Fully fleshed out inventive structures in hand, I progressed fully to the Translating phase.[3]

Translating

The Translating phase, so called because the creator is 'translating' their imagined text from mental representations to words (and other communication signs) on paper/screen (Flower & Hayes, 1981), is what most think of when we think of writing or creating. For *NW4T*, as for many born-digital multimedia works, this phase alternated between writing the fictional narrative, coding the underlying hypertext structure and mechanics and finding and manipulating the (relatively few) images incorporated into the IDN (cf. Skains, 2019b). The first element, writing the narrative, was relatively straightforward: Twine's back-end is specifically designed to enable the hypertext creator to visually structure their narrative, and to easily input text and links. As I now have an established narrative design practice in IDNs, I had already identified (in the Planning phase) various choices the reader-player could make, and so it was simple to craft the six endings and the narrative paths leading to them. Similarly, the last element, image editing, is an established part of my IDN practice; though it can be time-consuming, at this stage it does not present significant cognitive detours in the creative process.

Conversely, coding *NW4T* for certain communication and research purposes presented the steepest cognitive hills to climb. Two of these hills were based on narrative goals, and two were based on research necessities: (1) enabling the reader-player to customise the character names and pronouns to enhance the personalisation of the story for greater emotional affect; (2) designing subtly different endings depending on *all* the reader-player choices in the narrative, not merely the ones that lead to the final six alternative narrative paths; (3) recording reader-player gameplay for the purposes of project analysis and (4) requiring reader-players to log in to the game to track their gameplay to their participant identifier. Allowing player input that is then reflected throughout the narrative, such as customising character names, is a documented function of Twine, created using variables and conditions; Twine's popularity amongst marginalised communities such as LGBTQIA+ game designers means that several tutorials exist for this function as well as

---

3 Which was painfully poor—poetry is *not* my skillset.

customising pronouns throughout. Twine's in-built functionality using variables and scoring again enabled the different endings.[4]

Where Twine did *not* have native functionality, however, was in recording the gameplay. I had discovered this gap on a previous project (Skains, 2016; van der Bom, Skains, Bell & Ensslin, 2021): because Twine outputs as one single html file and thus *page*, it is impossible to use extant mouse- and click-tracking software to record reader-player activity. The record maps to the single html page, rather than all the separate passages of Twine, becoming useless. Twine creators are nothing if not resourceful, however, and between that previous project and YCO2, a couple had tackled this particular mountain and posted their code, using Google Sheets and Scripts in combination with Twine's variables and JavaScript functionality (Cox, 2018; Stewart, 2018). Even then, neither tutorial worked on its own, so I tweaked and combined them until the process worked as necessary.

The log-in page should have been a fairly straightforward coding process: providing an array of accepted values and then checking against that when the user enters them, using JavaScript. And indeed this worked fine, until an Intrusion of the Actual disrupted the process. On my previous projects, we had assigned participants their identifiers (usually simply numbers in a sequence: 001, 002, 003, etc.). My project partners, however, led the first YCO2 workshop and instituted a different identification system, wherein the student participants created their *own* identifier; I could not predict what these identifiers would be, and thus could not pre-set the identifier array.[5] What I would need was a database outside of the Twine game that the entire team could edit to enable users to have near-instant access to the IDN (for example, for students who weren't present in a previous workshop, to enter their identifier and enable them to read-play *NW4T* in a subsequent workshop). I assumed that since Twine could populate a Google Sheet to record gameplay, surely it could do the reverse and ping a Google Sheet to check for valid log-in identifiers. I used

---

4 *NW4T* includes 'macro-choices' and 'micro-choices' throughout the narrative. Macro-choices have an effect on the narrative pathway, such as choosing whether to go to school or to skive off with friends; these lead to one of six possible macro-endings. Micro-choices, such as whether to have a salad or a burger for lunch, do not affect the narrative pathway; collectively, however, they add up to a profile for the reader-player that ranges from 'environmental activist' to 'environmental destroyer'. A paragraph in each macro-ending passage changes depending on the reader-player's final profile of micro-choices.

5 Using a set equation: first three letters of their street name, first three letters of their mother's given name and the two-digit date (day only) of their birthday. All credit to my co-investigators—this was a necessary system for workshops that would be spread across months, as the participants never would have kept track of their identifiers otherwise.

Dan Cox's 2018 tutorial here again, every which way I could, but it would not function. After four days of working on it (some of them 12+ hour days), I broke down and, for the first time in my then 11-year career as a digital designer, posted a request to the various Twine forums (roseslug, 2018a, 2018b). Despite several responses, however, none hit upon the solution. I tweaked and tested it for another four days or so, before finally stumbling on the answer, which was a simple change in the Sheetrock JavaScript 'call-back' function. This function pulls the array from the Google Sheet to the Twine game; in Cox's example the specific code is 'State.variables.response = response.rows[0]'. In the end, the simple fix was to change 'rows[0]' to 'html'.[6]

These were not the only coding challenges I resolved in *NW4T* (I also worked quite hard to use CSS, html and Twine macros to create animated text message streams and pop-up passages); they were, however, the trickiest, and the ones that waylaid my translation process the most. The coding took all of February 2018 to sort, more than half of the two-month drafting period (Planning and Translating) from 3 Jan 2018 to 3 Mar 2018.[7]

Comparatively, *OAN* was refreshingly simple. Because I had already sorted and stored instructions and coding for recording gameplay and implementing log-in screens in *NW4T*, I used the same scripts with no problems. *OAN* didn't require custom character names/pronouns, nor did it incorporate any animation. A few elements of the visual design were briefly challenging (mostly a fade-in-out transition when the reader-player enters Oz), again due to Twine's single-html-page nature, but they were resolved in hours rather than days or weeks. The time necessary to complete *OAN* had little to do with the IDN itself, and more to do with Intrusions of the Actual in the form of personal events preventing the work (from job stresses to the pandemic). Nonetheless, it eventually wrapped up and moved to Review.

---

6 In retrospect, it seems an obvious solution, especially since Cox's comments in the script suggest the change for various sources. Without in-depth knowledge of JavaScript and how it interacts with Google Sheets and html documents, however, I wasn't able to make that connection. I rest a bit easier knowing that no one else on the message boards made that connection either, though their knowledge so exceeded mine that I didn't even know how to implement some of their suggestions. For instance, I'd love to be able to tell you why 'html' worked when 'rows[0]' didn't; sadly, I cannot, because I have no idea.

7 This isn't to imply I worked on this IDN full-time during this period. I was a full-time, active teaching academic during this period, though I did have an extra day of teaching buy-out from the project grant. So approximately 1–2 days per week for two months were spent drafting *NW4T*, until I unfortunately tipped a full water bottle of rose and pomegranate cordial over my laptop and murdered it dead. Luckily the text was both backed up and mostly finished.

Reviewing

The Reviewing phase occurs as the creator evaluates the text (either text-so-far, or the completed text) for how well it meets the project goals, responding to the rhetorical problem, and revises the text according to this evaluation. It is certainly oversimplifying the cognitive process to separate the phases in this way—after all, I would not have known my various scripts were failing to function if I was not engaged in a translate-review cycle in testing them. Linda Flower and John R. Hayes are careful to point out that their cognitive model, upon which the Practitioner Model is based, is *not* a hierarchical model with discrete and sequential steps (1981); rather, the stages or phases are teased out of the ongoing and overlapping cognitive processes for the purposes of analysis and discussion. Thus Reviewing occurs on both a micro scale, as the creator writes, monitors and evaluates the text-so-far on a word/sentence/section level, and on a macro scale, as the creator evaluates and revises completed versions of the finished text.

For YCO2 and IS this process was necessarily expanded, overlapping Intrusions of the Actual. Digital texts (e.g. websites, computer games, and IDNs) benefit from both *alpha-* (limited to the creator/s) and *beta-* (limited to a segment of the public) testing, in order to gather feedback not only on the aesthetics of the narrative media but also to evaluate the user interface and ferret out bugs (errors) in the code.[8] Alpha-testing, as noted, occurs periodically through the Translating phase on a micro-level, though final macro-level alpha-testing for these projects also extended to testing the IDNs on as many web browsers and mobile devices as I had access to. This was a relatively smooth process, as the micro-testing had exposed most of the issues.

Beta-testing extended to both known helpful testers (family, fellow digital creators) as well as the co-investigators on each project. This enabled an interesting dichotomy of beta-testers: those who were familiar with IDNs, and those who weren't. The former were able to speak to more specific design elements, such as game mechanics, design bugs, replayability and compliance or confusion with known IDN conventions. They typically played the IDN multiple times, expecting different outcomes, knowing they were testing multiple narrative paths. In contrast, the latter group were less useful on a technical or narrative level, as they rarely played through more than once (indicating a lack of

---

8 Obviously, this cycle of review and feedback is not limited to digital texts. Creative writing is workshopped, films are test-screened, plays are rehearsed. Interestingly, as fan fiction has expanded thanks to digital interfaces, the community has taken up the tech-term 'beta-readers' in place of the previous 'workshoppers'.

familiarity with IDN conventions); they were more helpful, however, in providing feedback from an inexperienced reader-player's perspective. Almost all fed back that they thought there should be 'more instructions'; yet when asked if the lack of instructions prevented them from playing or enjoying the IDN, all responded in the negative. Few IDNs of this nature provide any sort of instructions, similar to the lack thereof for today's smartphones; the designer's goal is to create a product that is 'intuitive', where the user can learn to use the interface in a playful manner, rather than plodding through instructions (that most don't read anyway). It was important for me to accept this feedback in this context, and to ask the follow-up question to determine if the IDNs really did need a layer of instructions, or if the testers were communicating their initial lack of familiarity with the form. If they responded that the lack of instructions kept them from playing or enjoying the text, I would have returned to the IDNs and revised them, either with instructions or a greater level of intuitive play in the first few passages. As no one gave this response, however, I was satisfied that the IDNs were intuitively communicating their mechanics, and made no changes. Indeed, none of NW4T's reader-players (teenagers, reading the more mechanically challenging of the two IDNs) indicated any confusion about how to interact with or progress the text.

Intrusions of the Actual

As already noted, Intrusions of the Actual have been significant on both of these projects, both serendipitous and disruptive. *Serendipity* in this model refers to the specific confluence of cognitive activity and external stimulation that sparks a creative connection leading to a productive outcome (Makri & Blandford, 2012a, 2012b). These projects provided spaces in which serendipitous connections could be made not only between the creator's existing knowledge and external factors but also between collaborators as we exchanged knowledge from our different areas of specialty. In a sense, we each provided a significant amount of external stimulation to spark connections within each of our cognitive contexts, increasing idea and knowledge connections in an exemplar of collaborative working.

Perhaps the best example of this for me was on the Infectious Storytelling project. As described in Chapter 3, the project was the result of various collaborative exercises between researchers in highly variable subject areas. In this case, a microbiologist (Hayhurst), a Romanticist (Casaliggi), a film scholar (Woodward) and I found we had mutual interests in plagues, infectious diseases and how they spread, how they affected society and how we responded to them in our cultural arts. Hayhurst pulled us together into a direction: the problem of AMR. As we began to brainstorm a potential project using the grade-school template of markers and flip-chart paper, it was clear that, given our assorted research areas, we would have an enormous domain to cover if we kept our focus to 'infectious disease'; we needed a particular case study disease. I looked around the table, and the research specialisms on display combined with the enthusiastic amateur knowledge I have of infectious disease (supplemented by a degree in agriculture and years of experience in veterinary practices and caring for livestock) sparked a light bulb—TB. I could almost hear the PING! as the idea connected in my mind. From my literary studies, I knew TB (or 'consumption') was rampant in the Romantic era, and many well-known artists suffered publicly from it. I knew from my science background and popular science reading that medical science and the development of vaccines and antibiotics occurred in a parallel time frame to the development of film, and that 'outbreak' is almost classifiable as a film genre. I also knew that TB is not, as most media portray it, merely a disease of yesteryear, and that it is in fact a re-emerging epidemic thanks to our central research premise: AMR. In retrospect, it seems the most natural choice, particularly once we uncovered details about the treatment programmes in Wales, the Romantic promotion of consumption as a source of creative genius, and the use of film in the mid-twentieth century to promote information about disease vectors and antibiotic campaigns. At the time, however, it was a tenuous, serendipitous connection.

Collaboration and the accompanying serendipitous connections fostered some of the most creative aspects of the project IDNs. *OAN*'s Tin Woodman is a poet (inspired by his canonical backstory as a hopeless romantic), reciting lines from John Keats' (1820) *Ode to a Nightingale* theorised (Tagore, 2000) to refer to his TB-related suffering; I drew this from Casaliggi's background research into Romantic artists and perceptions of TB in the Romantic era. Our purposeful narrative approach to our health communication message was inspired by the not-so-inspiring Public Health England's 'Antibiotic Guardian' campaign (PHE, 2014), as analysed by Woodward, which was of high production value but dubious impact. *YCO2*'s first workshop on carbon footprints, designed by Rudd, inspired many of the micro-choices in *NW4T*: transportation choices, food choices and leisure time choices.

Teamwork inspires serendipitous connections, but it can also provide many distractions in the form of interruptions or changes of direction. *YCO2*'s original project timeline scheduled *NW4T* to be finished by early April 2018; we made an additional school connection, however, which meant running all the workshops in a compressed 'STEM week' a month earlier than planned. This acceleration led to the strict three-month drafting time for *NW4T*, and did not enable wider team collaboration in the creative process. On the other hand, the more open timeline of the IS project meant that my personal circumstances repeatedly pushed back the completion of *OAN*, which I'm sure provided a source of frustration for my team members.

Working with others can both aid and hinder the creative process. For instance, we identified a need with YCO2 to get students over the 'I don't know what to write about' hump they often experienced at the start of workshop 3; as teaching moved to more online needs in 2020, we surmised a mini-Twine game might help. Rudd, the YCO2 PI, came to me with this request, but we struggled in terms of communicating the rhetorical problem (and thus the solution) to one another. Further, she jumped in to trying to show me what she wanted with the coding by *doing* some coding. On the one hand, she very much enjoyed playing with the Twine interface and getting to know that aspect of our project. On the other, it led us for several hours in the wrong direction, as her long-term memory regarding the tech was insufficient to create the effect she aimed for; as a result, her goal was completely obscured from the person who had the long-term memory/expertise to effect it (me). This was easily mended with a phone call, but this issue can occur on much larger scales in collaborative projects, with much more dissatisfying results.

Monitor

Monitoring during project creation is generally extremely helpful to avoid texts that miss the mark responding to the rhetorical problem or meeting the goals set in Planning. Like the Reviewing Phase, it has both micro- and macrolevels, as the creator monitors each word and image for its usefulness to the text as well as the overall text-so-far for its suitability for the rhetorical problem. In the above example, it helped to resolve Rudd's and my communication problem, as I monitored and reviewed her coding and determined it was not serving the rhetorical problem as I understood it. This can occur in both individual and team efforts.

Monitoring is also enhanced by repeated experience in the Practitioner Model, or more generally practiced professional communication strategies. Professional communicators (of which creative writers and narrative designers are certainly included) constantly engage in this cognitive process, though individual creators can be very likely to alter the rhetorical problem if they find their Fictional World has strayed into unexpected territory. For example, I have started with a rhetorical problem such as 'complete a short story' then changed course midway through as I discover the narrative is extremely visual and better suited as a script; the rhetorical problem then becomes 'complete a script'. For individual creative endeavours, that doesn't pose much difficulty; in collaborative projects with highly defined outcomes for highly specific communicative purposes, the rhetorical problem has far less flexibility. Thus monitoring—micro- and macro-, as individual and team members—becomes far more important. It helps keep the project on track and on (rhetorical) task.

CONCLUSIONS AND RECOMMENDATIONS

Writing to educational spec presents significant challenges compared to straight-forward artistic creation, but a considered and prepared approach can be effective. For the creative practitioner, the challenges present as: Long-Term Memory gaps that must be filled prior to creating; greater fixity in the project's Rhetorical Problem that can both hamper and inspire creativity; a more complex and concrete Planning phase that ensures the creative work aligns with project goals; a more rigid Translating phase, less able to follow whims of the Fictional World created, and more susceptible to Intrusions of the Actual; and Disruptions from those intrusions in the form of anything from changes in timetables to shifting team goals. The creator may also experience some frustration from these effects on their established creation process, as well as from potential shifts in creative focus on aesthetics to more practical considerations in order to meet the needs of the project and its audience.

Conversely, collaboration on works for specific educational and communication needs can significantly enhance the creative process and the work that results, as long as the creator is accepting and adaptable to the various challenges noted. The ability to accept, analyse and utilise feedback on the creative work is heightened when constructing a creative work in a group for a specific purpose. Perhaps the most important aspect of that for the experienced writer/designer, for whom simple professional practice should already include accepting feedback and making changes, is the *analysis* aspect. Health and science communication teams are interdisciplinary. The scientist on the team may not be able to thoroughly understand or articulate the reasoning behind use of first-person narrative perspective versus second-person narrative perspective, any more than the project practitioner could detail why lab chemicals should *not* be stored in alphabetical order. Thus the feedback the practitioner receives may not be as targeted and nuanced with respect to the rhetorical problem, creative techniques and even technological functionality as they are used to from more familiar creative reviewers and editors. Regarding the creative work and its applicability in the research and/or communication project, the practitioner may fulfil dual roles as project collaborator *and* service provider, considering and responding to team members' feedback as though they are non-expert *clients*. It can be a very frustrating process to adjust a creative work according to feedback from those outside of creative domains (see: common narratives about Holly-wood studio execs 'ruining' films, abstract art amateurs claiming 'my kid could do that', and business clients attempting to dictate font choice and visual layout to web designers). The passion and ego that drive creators' artistic visions must compromise with the communicative goals of the project and its team; perhaps this challenge is part of why these art-science collaborations have been relatively few. With deeper understanding of the creative cognitive challenges on these projects, as well as the interdisciplinary working dynamics of the teams (discussed in the next chapter), hopefully more of these projects will emerge to all of our edification.

5

BRIDGING RESEARCH SILOS: APPROACHES TO ARTS-SCIENCE COLLABORATION

INTRODUCTION

The lead author once presented a poster at a conference for the 'empirical' study of literature. As she'd done an ethnographic study of writers experimenting with digital composition, she thought she'd encounter a lot of other scholars in the domains of composition study, literary criticism and creative writing. In reality, she was almost entirely alone as a humanities scholar at this conference; most of the attendees and presenters were psychologists studying how narrative affects the brain and how we understand ourselves and others through reading. The experience was both fascinating and enlightening, as she discovered an entire field of researchers were studying the same things hers did, but in completely different ways using completely different methodologies, terminologies and understandings of their subjects. She became something of a minor celebrity at the conference, as in each session she couldn't help but raise her hand and deliver insights from *the other side*, and how narratology studies or gender studies or adaptation studies could enhance the researchers' approaches (and, to a certain extent, how their ignorance of much of this discourse could confound their work). What's more, the organisation's general meeting revealed one reason for their isolation from the more humanities-oriented domains related to literary study: they couldn't get literary scholars to work with them.[1]

---

1 An experience the lead author could commiserate with.

That conference is a microcosm example of what happens all across research, as the breadth of existing knowledge is simply too vast to manage in any one individual.

> *Humanity's ever-growing store of knowledge, and the fact that each person is bestowed with a unique set of aptitudes, left most scholars and artists stranded in ever-shrinking islands of competence.*
>
> *(Nissani, 1997, p. 202)*

To earn our ticket into academia, we must complete a years' long, usually independent research project demonstrating an original contribution to knowledge, which increasingly results in incredibly deep dives into the minute particulars of a research area. We emerge from this gauntlet an expert in a niche of one, and supposedly this readies us for life as academic researchers.

Increasingly, however, academia and its associated funders are waking up to the difficulties inherent in strict disciplinarianism, and calling for interdisciplinary research. Questions remain, however, as to what 'interdisciplinarity' truly means; it could refer to a project incorporating researchers from a variety of disciplines, an individual researcher drawing on multiple methodological approaches, research outcomes conveyed across various pertinent disciplines and so forth. How do we, trained along singular, remarkably refined and differentiated tracks, *do* interdisciplinary research?

This chapter discusses approaches to interdisciplinary research, and examines the two case studies presented in this book for their particular methods.[2] We offer here the history of how the projects came to be, and insights into the relationships between the researchers with one another and with the projects themselves. Of particular note is the 'wide interdisciplinarity' (Kelly, 1996) employed on these projects, in that they stretched the gamut from science to arts; there are relatively few in-depth descriptions of inter-disciplinarity at work, and far fewer regarding science-arts projects (Klein, 1990). Our experiences show that despite the disparities inherent in arts-science disciplines, a well-researched and well-conducted training pro-gramme combined with alignment of researcher values provide a strong foundation for productive research on wide interdisciplinary projects. The sections that follow offer an overview of interdisciplinarity and its barriers and benefits, a more targeted discussion of arts-science intermingling in research

---

2 Note that the bulk of the research into interdisciplinary research occurred after the projects had already been conducted, in large part. The comparative analysis in this chapter is post-textual.

projects, and an examination of both the You and CO_2 and Infectious Storytelling projects with regard to their interdisciplinary workings.

UNDERSTANDING INTERDISCIPLINARITY IN RESEARCH

Our current system of codification and specialization emerged as part of the institutionalization of knowledge, in a nineteenth-century rage of classification, taxonomies and disciplines (Nissani, 1997). Prior to that, generalism was prized: universities were so called because they promoted 'universal' learning from philosophy to mathematics, and the notion of a polymath or (annoyingly gendered) 'Renaissance Man' reigned. Shifting to an academic structure of specialisms, however, enabled greater depth of research in single areas. Of course, this disciplinary approach became entrenched in research institutional structures for decades, until more contemporary gaps in knowledge exposed the weaknesses of this 'siloed' system (Chettiparamb, 2007; Klein, 1990; Stirling, 2014).

As a current buzzword for institutions and funders, it can be difficult for researchers to suss out just what the term 'interdisciplinary' means to its various employers, as well as what types of projects to apply it to. This can be understandably frustrating when a researcher is seeking support for a specific, disciplinary research project, and can only find funding opportunities that require multiple researchers and disciplines, as funders seek to promote the current zeitgeist. Similarly, the term 'interdisciplinary' is used in many different ways. Often, it refers only to science disciplines: the bulk of the literature about interdisciplinary research is restricted to science-based projects and disciplines. It is also used in projects more accurately described as multi-disciplinary, which 'is about more effectively "joining up" the contributions of different disciplines... with each aspect addressed in a neatly differentiated fashion' (Stirling, 2014, n.p.).

For the purposes of this discussion, we are differentiating multi-disciplinarity from interdisciplinarity. The former is, as Andy Stirling notes, the use of differentiated disciplines toward one goal, like members of an *Ocean's Eleven* con job, not really knowing how their colleagues get their portion accomplished, only that they do (and on time, without getting caught out). Interdisciplinarity, on the other hand, '*combines* components of two or more disciplines in the search or creation of new knowledge, operations, or artistic expressions' (Nissani, 1997, p. 203, emphasis ours; cf. Barry, Born, & Weszkalnys, 2008), particularly to address problems or questions that cannot be answered using single or siloed approaches (Marzano, Carss, & Bell, 2006, p. 186). Such problems include issues of a global nature, and those addressing human needs, discussed more below.

Despite the current emphasis from funders on interdisciplinary projects and optimistic notes on how interdisciplinarity can resolve difficult problems, however, it would be a mistake to assume academia should shift entirely to interdisciplinary approaches for all subjects and projects. On the contrary, interdisciplinarity is not inherently better than disciplinary methods and knowledge; like any method, however, it needs to be applied to the appropriate questions and projects. Julie Thompson Klein notes that 'Interdisciplinarity and specialization are parallel, mutually reinforcing strategies' (1990, p. 7), and that the relationship between the two creates a complex productive tension. Interdisciplinary projects can be successful— more successful than other approaches, depending on the problem being addressed—specifically *because* of the disparate disciplines coming together, bringing unique knowledge sets and perspectives to the issue at hand. Each discipline plays an appropriate and necessary role (Bauer, 1990, p. 113), like team members in different playing positions on the field, working in complementary ways to accomplish a unified goal.

That joint goal is the key to interdisciplinary research. Every research project need not be interdisciplinary; certain questions call for specialised approaches, and forcing an interdisciplinary stamp on them confounds the process. Other questions, however, particularly those of global and/or novel problems, call for a multifaceted approach that interdisciplinarity is ideally suited to. '[I]nterdisciplinarity is commonly identified with problem-solving in response to new problems or objects that, it is believed, lie beyond the frame of existing disciplines' (Barry et al., 2008, pp. 29–30). In particular, interdisciplinary approaches are applicable for projects 'involved in the development and sustenance of self-hood' (Kelly, 1996, p. 100); problems occurring in 'focal points where disciplinary social worlds intersect' (Klein, 2000, p. 13); complex questions, broad issues and those beyond the scope of any one discipline (Klein, 1990, p. 11); and developing and modifying existing disciplines with perspectives, processes and frameworks from others (Bechtel in Klein, 2000, p. 6).

Specifically, interdisciplinarity introduces numerous benefits to research projects to address these wider-scope questions. Moti Nissani discusses ten 'cheers' or benefits to interdisciplinary working that strictly disciplinarian approaches often lack (1997). Outsiders' perspectives on issues often lead to creative breakthroughs, and any given interdisciplinary team has numerous outside perspectives built in (Bromme, 2000; cf. Hollingsworth

& Hollingsworth, 2000). Cross-disciplinary oversights, such as those described in the opening to this chapter, where multiple disciplines are labouring in the same area with no awareness of one another, can be minimised; likewise, work that falls into the 'gaps' between disciplines can be highlighted. Interdisciplinarity also encourages flexibility in research, allowing academics to explore new and stimulating areas rather than falling into the trap of disciplinary tunnel vision leading to diminishing returns. And perhaps most importantly, interdisciplinarity permits and encourages a 'unity of knowledge', allowing us to piece together the individual puzzle pieces of disciplinary knowledge into a whole picture. This holistic capacity enables interdisciplinary teams to address more complex or practical problems, particularly those related to humans, society and social change.

There are certainly, however, many barriers to interdisciplinary working, from institutional structures to individual personalities. Academic institutions are built upon a long tradition of disciplinarity (Weingart & Stehr, 2000), and often struggle to provide sufficient space and culture for interdisciplinary projects: 'as cynics have stated it: "The world has problems, but universities have departments"' (Brewer, 1999, p. 328). The current metrics in place for research evaluation reinforce these distinctions, urging more work in silos, despite broader calls for interdisciplinarity (Stirling, 2014); this leads to professional impediments for interdisciplinary researchers and projects, including diminished funding, hiring, promotion, status and publishing (Bracken [Née Bull] & Oughton, 2006; Brewer, 1999; Nissani, 1997). Where does one publish outcomes from a project where a creative writer and a chemist worked together on a cultural issue? Which funder should such a project approach—and risk being rejected by *all* because the arts funder sees it as science research, and the science funder as arts? For those subject to the UK 'Research Excellence Framework' metrics, which panel should these outcomes be reported to?

These institutional disciplinary splits serve as the foundation for the distances between individuals on interdisciplinary teams. Each member of the team has been trained in a discipline, often for decades (Bromme, 2000). They have been trained in specific methods, working in specific operational frameworks, with specific frames of reference and background knowledge (Brewer, 1999). They value certain outputs over others, and apply specific analytical approaches. They understand data in a way specific to these methods and outcomes. For a neuroscientist, data may be defined as spectroscopic readings of neurochemicals, processed by analytical software and compared to previous such datasets. For a psychologist, the data may be defined as responses to a series of images on a screen. Even in such apparently highly related domains, it can be difficult to bridge the gap between disciplinary training, understandings and approaches; the further apart disciplines become, the more difficult this bridging can be.

Communication between disciplines is a key aspect in this difficulty. '[D]eveloping a common language and introducing colleagues from other disciplines to one's own perspective are described as the key problems of interdisciplinary cooperation' (Bromme, 2000, p. 123). Each discipline develops its own epistemology, its own language or dialect (Bracken [Née Bull] & Oughton 2006; Brewer, 1999; Bromme, 2000); these disciplinary dialects enhance communication within the discipline, as they reference deep background knowledge frameworks, relations, concepts and expectations. Outside of the discipline, however, these dialects hamper communication, as outsiders struggle to parse the subtleties, connotations and contexts inherent in disciplinary dialogue. Indeed, many researchers have reported struggling to properly convey their approaches and values to those outside their disciplines (Brewer, 1999; Bromme, 2000; Marzano et al., 2006), therefore finding difficulty expressing their perspective on a problem or project.

Conversely, researchers and practitioners can be stymied by the opacity of background knowledge in other disciplines: 'the technical skills, understanding and experience required to operate within a domain can be opaque or intractable to non-specialists without adequate training' (MacLeod, 2018, p. 707; cf. Marzano et al., 2006). These difficulties are compounded by a false consensus effect that leads us to overestimate others' knowledge about our own disciplines, which results in a failure to properly address the gaps between our knowledge and theirs (Bromme, 2000). And yet another compounding effect is the occasional perceived *similarity* between concepts or terminology between disciplines, where two researchers believe they have a consensus on a topic, but are each talking about different things; for example, 'observation' means very different things to primatologists versus theoretical physicists, and a discussion where these differences are not made explicit could lead to significant gaps in understanding. These problems of opacity and consensus result in 'mismatched communication and uncertainty in interactions between collaborators which make it very difficult to coordinate practices in productive ways' (MacLeod, 2018, p. 709).

The creation of disciplinary research not only generates communication and understanding gaps, it also creates a host of cultural differences. Disciplines have distinct operational objectives: humanities disciplines are usually discourse-based in an ongoing dialogue concerned with questions about humanity, value and meaning, while 'hard' sciences seek unassailable facts that can be used to understand the universe and its material interactions. This 'subjective-objective gap' results in a fissure between these disciplines 'because scientific presuppositions render value questions meaningless, neither true nor false, and meaning questions otiose, lost in the causal complexity of physical systems' (Kelly, 1996, p. 97). These gaps between disciplinary values often lead to judgements about the quality or worth of other disciplines, including disagreements about quality standards for research claims and conclusions (Bracken [Née Bull] & Oughton 2006; MacLeod, 2018). Lack of respect for other disciplines, generated by lack of understanding and difficulties in communication, results in superior attitudes, defensiveness, frustration and resentment (Bracken [Née Bull] & Oughton 2006).

Finally, interdisciplinary projects can be impeded by individual differences between team members due to internalised cultural differences or simple personality clashes. Researchers often align their personal identity with their discipline, and can view alternative perspectives or identification of weaknesses in their discipline as an individual assault. They may have chosen their discipline because they prefer its ways of working (or, have been institutionalised to them), and thus resist the conceptual and methodological cognitive leaps necessary for truly interdisciplinary work (MacLeod, 2018). Rainer Bromme points out that 'the predictive value of interindividual trait differences for interindividual behavioural variance is relatively low' (2000, p. 117); because disciplinary cultures, training and dialects can be so implicit to an individual's understandings and approaches, general personality traits cannot predict how well someone will work in *interdisciplinary* circumstances.

Despite the noted benefits of interdisciplinary work from the varied perspectives of disparate disciplines, it can be precisely that variation that impedes cooperation on these projects. As noted above, most literature on interdisciplinary work focuses on that between scientific disciplines, or 'narrow interdisciplinarity' (Kelly, 1996); bridging the gaps between supposedly vastly different disciplines (or James S. Kelly's 'wide interdisciplinarity') such as science and arts can both ease these difficulties and enhance them. The following section looks more closely at these wider endeavours.

WIDE INTERDISCIPLINARITY: ARTS AND SCIENCES

Both You and CO_2 and Infectious Storytelling are examples of wide interdisciplinary projects, combining as they do arts and sciences approaches to practical questions of human need (Kelly, 1996, p. 99). Given the cavernous gap between training, communication and cultures of scientists versus humanists and/or artists, the barriers as discussed above can be even more present in these wide collaborations. On the other hand, sometimes this distance is beneficial, as it lessens the occasions when various team members *think* they're talking about the same thing, but actually have very different perspectives on the same topic—and can't recognise it enough to clarify matters. By navigating the cultural differences with regard to research approaches, wide interdisciplinary teams can tackle some of the more complex problems surrounding society, human behaviour and how we understand and interact with our world. To do this, interdisciplinary teams, and particularly leaders of these teams, can manage and adjust their workings in the following areas: team culture, communication between members and individual member personalities.

Member personalities is perhaps the least tractable and yet most likely indicator of team/project success. Michael Anbar identified four types of 'bridge' scientists (1973 in Chettiparamb, 2007, p. 29), and this typology still has merit: (1) *adventurers*, who achieve grounding and satisfaction in one discipline, and go looking for new challenges; (2) *migrators*, who are grounded in a discipline that has become obsolete, forcing them elsewhere; (3) *generalists*, who have superficial training in one or more areas and (4) *administrators*, managers or sales staff who must bridge, but have no training or preparation to do so.[3] Adventurers, seeking as they are new challenges, will contribute the most to innovation and creativity, followed by the more reluctant migrators; generalists are useful in secondary purposes (marketing, organisation) but their lack of deep grounding in any one discipline will actually be detrimental to true interdisciplinary working. Administrators, as we have labeled them, are the most likely obstacles to interdisciplinary work.

While Anbar did not break down quite what makes someone an Adventurer, some likely traits are a strong ego (to weather the inevitable misunderstandings and potential disrespect for their discipline and research), initiative and/or assertiveness (to convey their/their discipline's importance to the project), a broad education (which enables at least some understanding of other disciplines) and a 'sense of dissatisfaction with monodisciplinary constraints' (Klein, 1990, p. 183). Klein notes that divergent thinkers are more likely to succeed than convergent, and that the ability to differentiate, clarify, compare, reconcile and synthesise information and data from disparate sources and in different frameworks is also a key indicator of a successful interdisciplinary personality. Place these qualities in a researcher interested in

---

3 Anbar identified the types; we applied the specific labels 'adventurer', 'migrator', 'generalist' and 'administrator'.

problems of social and societal importance, and a picture of an interdisciplinary researcher emerges.

Regardless of how 'personality-perfect' each member in an interdisciplinary team may be, the team culture must form around trust, receptiveness and a willingness to learn (Marzano et al., 2006). An interdisciplinary project will include methods, analysis and even publication strategies familiar to some but not to others. There will be stages in which one team member must instruct the others in the rudimentary elements of their discipline, which can feel like pandering to the teacher and patronising to the rest, if not handled in a culture of trust and equal exchange. This receptiveness also softens the messages about the weaknesses of one's discipline, which are sure to emerge in interdisciplinary projects; if a researcher is to contribute to a project that combines disciplines for increased strength, then they must be capable of acknowledging where their approach would weaken the project structure. Interdisciplinary working requires a degree of humility, if not necessarily in one's own work, then toward one's discipline as a whole, in order to co-create a common ground for the project team.

Particularly on *wide* interdisciplinary projects, it is important that all team members value one another and one another's contributions to research equally. Quite often, arts/humanities (hereafter: humanities/humanists) are treated as a communication device for the sciences: a 'true' researcher (the scientist) creates data, analysis and conclusions, and the humanist puts it in a form more palatable for the public (Barry et al., 2008, p. 29). This demonstrates an unequal partnership, one that is dismissive of the humanists' research, methods and insights. Andrew Barry, Georgina Born and Gisa Wezkalnys note that even though this attitude is one of the justifications for funding arts-science research in the UK, the reverse can actually be true, as scientists provide resources and equipment for artistic projects (Barry et al., 2008). In either case, these do not fit the bill of truly interdisciplinary work—rather, they are more on the level of services rendered.

What creates and sustains a healthy interdisciplinary culture is open, frequent communication. The benefit of these projects is specifically the multiple perspectives and approaches the different disciplines bring; without proper communication, however, the team will find no common ground (or 'grounding') (Bromme, 2000). Dialects, formed as noted above from disciplinary uses distinct from ordinary usage, are tools of disciplinary cohesion but supreme opportunities for misunderstanding in mixed-discipline settings, particularly when terms cross disciplinary boundaries but their meanings do not. Grounding requires a system of shared common references; the same process of training and language-building that creates disciplines and their dialects can be used to create a grounded culture in interdisciplinary teams. In line with open communication in a culture of trust, team members must be willing to teach others about their disciplinary background knowledge, methods and approaches, and to learn such about other disciplines. In doing so, they may employ specific communication methods such as articulation and metaphor. Articulation calls for a deconstruction of each member's disciplinary knowledge to basic building blocks, and reconstruction in the team to form a common understanding of each (Bracken [Née Bull] & Oughton, 2006, pp. 377–378). Disciplinary pairing, or including two team members from one (or similar) disciplines, can aid communication in that it doubles the number of ways their discipline can be explained to other team members (ibid., p. 380). Metaphor, as one of the key learning strategies for all humans, is another important method of communication (ibid.; Bromme, 2000). They create common ground by relating a specific disciplinary concept to a more general one.

Despite even the best intentions, the task of communicating between disciplines can be difficult. In any given situation, we as communicators make assumptions—whether conscious or not—about what our audience knows and understands, and we adjust our messages to be appropriate. Unfortunately, as researchers in a group with other researchers, we may overestimate what our team knows about our specific discipline (Bromme, 2000). Individual personality traits and disciplinary insecurities also play a key role here, as well as external cues. The linguistic phrase 'it's not rocket science' and the cultural attitude that 'math is hard', for instance, may influence an artist, historian or sociologist to subconsciously assume an aeronautical engineer or mathematician is smarter than they, and lead to failure to thoroughly express the basics of their own highly specialised (and thus dialectic) disciplines. Imposter syndrome can have the same effect. Researchers in certain disciplines may have internalised biases about other disciplines and their methods, leading to inadvertent disrespect and resultant loss of trust in their exchanges. Thus, frequent meetings, preferably in-person to gain the full spectrum of body language and non-verbal cues, with clear agendas and an atmosphere open to even the most basic questions and disciplinary vulnerability, are recommended to foster trust and grounding (Marzano et al., 2006).

For an interdisciplinary project to succeed, then, at least on the basic level of 'keeping the band together' and working toward a common research goal, several elements must develop. The team members must have shared values, even if only at the level of project goals. It may not be possible to share all values on a team, particularly as the disciplines involved become more disparate; in these cases, it is important that the team remain flexible, and open to including individual secondary goals to increase or supplement member investment in the project. Each member must have equality in the team: equal input and outcomes in terms of methodological input, background research and publication. Only by mutual agreement of the team should this dynamic change (such as for personal circumstances, part-time careers, etc.). Perhaps more importantly, each member should hold equal respect for every discipline involved in the project, trusting that everyone brings an advantage to the interdisciplinary table, even if one struggles to understand it. This mutual respect generates trust, flexibility and an openness to learning that facilitates communication. Communication itself is a cycle of trust and respect, as the team will need to develop a project dialect from the bases of disciplinary dialects present; this requires honesty from each member about their strengths and weaknesses, as well as those of their disciplines. It also requires a commitment to frequent and appropriate communication, and effective project management from the principal investigator or team leader.

You and CO_2 and Infectious Storytelling are wide interdisciplinary projects that have maintained strong team cultures since their inception in 2018. As we discuss in the next section, these dynamics were certainly not created by chance, though as team members we were often unaware of the processes of culture-forming even as we were participating in them.

WIDE INTERDISCIPLINARITY IN ACTION

The two project case studies in this text (You and CO_2, and Infectious Storytelling) both emerged from the Welsh Crucible (2013) nationwide training programme in 2018. The selection process for this programme, and the activities of the programme itself aligned the project teams for success—though as participant researchers, we were unaware of the eventual goals of many of the individual activities. The analysis of the programme and our projects as presented in this section is applied post-textually, as it were, at a point where we have achieved a greater understanding of interdisciplinarity and how interdisciplinary teams work. The successful working of both project teams is largely due to the tremendous preparation we received as part of the Crucible programme, from its selection process to its post-training support.

Welsh Crucible: Foundations

The Welsh Crucible is based on the NESTA 'Crucible' development programme (NESTA, 2018), an extensive suite of materials to aid third parties to run residential leadership programmes explicitly for the development of interdisciplinary research.[4] Each year, 30 researchers in Wales are selected to participate in a series of three 2-day residential workshops. Welsh Crucible is supported by a consortium of higher education institutions across Wales, with most participants coming from universities: Cardiff, Aberystwyth, Bangor, Swansea, Cardiff Metropolitan and the University of South Wales. Though early versions of Welsh Crucible were primarily concentrated on science researchers, by 2018 it had expanded to include the arts and humanities (who were, nonetheless, still a minority group).

The process of selection for participation is rather opaque, deliberately so, which became a source of mythology on the programme. Each institution has a 'Crucible Champion', someone who drums up applicants at the start of the year, conducting 'road shows' to market the Crucible to researchers in the institution, and helping applicants construct their proposals. The application itself is rather different from what most researchers are used to: the emphasis is not on the researcher's accomplishments, grant capture, awards or publications. Instead, the focus is on who the researcher is and what they want. The application asks about the applicant's research, in brief, but continues on to ask about the applicant's broader research interests, and interests outside of research entirely. According to the Crucible Champions, they select the top 30 participants each year, regardless of discipline, gender or career stage and/or accomplishments. What earns that moniker of 'top' is down, largely, to how open the participant is to new, collaborative and interdisciplinary working.

The programme consists of three 2-day residential workshops over the course of three months. All participants are required to attend all elements of the workshops, including staying in the workshop hotel, even when the residential is scheduled in their hometown. From arrival on Wednesday evening to departure on Friday afternoon, 'cruciblees' spend all of their time together, whether it's in workshops, seminars, visits to Welsh Senedd, collaborative activities and mealtimes including dinner and breakfast. While some of these 'extracurricular' sessions are used for networking (as the programme brings in outside interested parties, previous Crucible attendees, etc.), they are primarily used to build camaraderie and trust amongst the participants. After all, with 30 researchers from six universities and countless departments, it would be impossible for them to get to know one another during workshop and training

---

4 Unfortunately, the NESTA Crucible in a Box site has been deprecated, but can still be accessed on the Internet Archive at https://web.archive.org/web/20180829104756/http:// crucibleinabox.nesta.org.uk/.

time alone. This (initially) forced togetherness fosters a key aspect of inter-disciplinary working: trust.

The three residentials moved participants through three phases of training. The first residential was focused on introductions; 30 new people and their research areas are difficult to grasp when we would only see each other for a few hours over the course of three months. Activities included icebreakers, pecha kucha presentations on ourselves and our interests, and elevator pitches about our research. Lectures focused on communicating our research, both to other researchers and the public. The second residential focused much more extensively on exercises to bring our disparate disciplines together, such as a 'speed-dating' event and brainstorming groups, and on how different types of people can effectively work as a team. This involved some 'working person-ality' tests; while many of us scoffed at the accuracy and application of such tests, they nonetheless provided the opportunity for us to have open discus-sions about how we worked with people, and how we could improve those relations. The final residential focused on career development for the future, and provided numerous activities for the developing research groups to come together and plan their projects.

The early phase of our particular Crucible programme highlights the dif-ficulties wide interdisciplinary project teams often face. For one, scientists far outnumbered humanities/arts researchers in our group. Much of the initial casual discourse (at least as we can perceive in retrospect) indicated that the scientists had no idea how or why they would ever collaborate with the humanists. This attitude was compounded somewhat by a benignly skewed attitude from the facilitator, who predicted that the computer scientist with a specialism in artificial intelligence would be the belle of the ball, and indicated indirectly that the humanists in the room would need special help under-standing everyone else, necessitating a 'canary' who could sing when the sci-entific discourse got 'too hard'. She was off-base on both accounts; many of the scientists eventually saw more benefit to working with humanists than algorithms, and it turned out the scientists more often needed primers on humanities topics than the other way around.

We also encountered the notion that the humanists would simply be expected to serve as mouthpieces for scientific research. One scientist pro-claimed excitement that they could do the science, and then the humanist in the group could put it into a form more palatable for public consumption. The benefit of having a humanist as a programme leader shone brightly at this point, as he quickly pointed out that humanists are not merely tools, and that they also had research objectives that should be considered. Subsequent activities on the programme, in particular a 'speed-dating' event, encouraged more of a two-way dialogue between individual participants, and enabled each to understand the personal and research interests of the others. By the end of the second residential, groups with shared interests—some of which were new to each participant, emerging from mutual conversations and objectives—had begun to form.

How Our Projects Bridged Disciplines

By the third and final residential, each project group had begun to draft an application for seed funding from the Crucible programme. In addition to cementing group formation, this application process pushed each group to start forming many of the necessary elements for interdisciplinary working, as described above: shared values, openness to individual secondary goals, team member equality, equal respect for each discipline, shared project dialect and frequent and appropriate communication.

For both You and CO_2 and Infectious Storytelling, the *shared values* are what drew the groups together to begin with, primarily in the second residential. A key researcher in each group identified a strong personal value: Dr Rudd expressed a deep passion for combating climate change, Dr Hayhurst for combating antimicrobial resistance. Each drew collaborators into their circles based on these interests, and at this stage of the crucible programme we were all very prepared to step outside our disciplinary boundaries to consider how we could work together toward these shared objectives. Thus both projects formed based on shared values, which provided a strong basis for working together.

On each project, however, we noted that we were all individuals at the end of the day, and would need to be able to show—in the UK's increasingly stilted and metric-based evaluation system—our contributions to the research and how the projects were advancing our own portfolios. This task is more of a challenge than it seems. For instance, in Dr Rudd's chemical engineering environment, the You and CO_2 project as a whole would not enhance her metrics or her career, even in terms of impact as her department defined it. Dr Casaliggi's area of Romantic art and literature could easily find itself left out of the research outcomes considering that antimicrobial resistance is certainly a contemporary issue. Dr Skains could have found herself constructing IDNs for the projects based on instructions and information she had no investment in. Thus we built into each application an outcome or opportunity for each team member to 'show' progress or success in their individual disciplines, if the project as a whole was not suitable for that purpose. We elected Dr Rudd as PI in the You and CO_2 project not only because of her passion for the topic but also because the role would be more beneficial to her experience and career as a postdoc than to the rest of the team, who were all Senior Lecturers at the time. We selected tuberculosis as a case study on Infectious Storytelling so that each team member had a valid and useful avenue of research to pursue, whether in Romantic literature, film, games or medical history.

On both projects we have worked to ensure all team members have equality. This does not necessarily mean that all contributions are equitable; by agreement, some members contribute less, and some more, depending on their investment in the project and their other priorities. This flexibility is absolutely necessary in any team. Over the two years of these projects, we have had babies born, institutions changed, career tracks changed, houses moved, fellowships gained, personal struggles and health issues and, of course, a pandemic. At various stages we have all had to make accommodations for one another. One team member on the You and CO_2 project found their circumstances so changed that they had to step down, an unfortunate occurrence that was compounded by difficulties in communication with them.

Contributing to this equality is the mutual respect we have all formed for one another's disciplines. Our experiences on the crucible programme aided us a great deal here, as they enabled us the opportunity to really see, hear and talk to researchers in a vast array of disciplines, and to appreciate that all of our areas are equally valid and of intellectual rigour. A great deal of trust is necessary here as well: we must each trust that our team members can carry their load in their own areas. Part of that comes from communication, and in acknowledging where we don't understand something, where our disciplinary methods are weak, and in striving to learn—at least on a rudimentary level—the dialects, methods, strengths and weaknesses of our team members' disciplines.

From this foundation of trust and respect develops a project dialect, a way of talking to one another that is not ensconced in the deep-learning of a single discipline, but the wider understandings of the team. On the Infectious Storytelling project, choosing TB as a case study enabled us all to dive into a topic we hadn't really studied before, and we came together over our discoveries here. We watched films that portrayed its science and its cure, marveled at the poets who seemed to seek out the contagion and very much delighted in the unique experience—for all of us—of doing archival research on the national TB response in Wales. We all got up to speed on the medical history, on John Keats and what the Long Nineteenth Century is, TB's representation in film and games, and how IDNs work. Not only the shared value of combating antimicrobial resistance, but TB as a case study provided a mutual dialect to enable us to overcome our disciplinary, linguistic silos.

Finally, interdisciplinary teams, particularly those that are also interinstitutional, must commit to frequent and appropriate communication and project management. Given how well the Welsh Crucible prepared and placed us for these collaborative projects, both teams achieved the previous interdisciplinary elements fairly easily. This final aspect, however, is what can sustain a group, or break it forever. The two projects had different approaches affected by various circumstances. You and CO_2 benefitted greatly from its PI, Dr Rudd, and her enthusiasm and dedication. When one crucial team member trailed off the project and eventually dropped out, she wasted no time in securing another, Dr Ross, who benefitted the project greatly. Dr Rudd's communication is frequent, and often individual, as she formed individual relationships with each team member based on their contributions to the project as well as their mutual interests. This has enabled the project to develop at breakneck speed, with several further successful funding applications (and several unsuccessful, of course).

With Dr Skains at the helm for Infectious Storytelling, this project progressed much more slowly, as the period of 2018–2020 ushered in a great deal of professional upheaval for her, with her initial university phasing out her research and teaching area in a period of mass redundancies, leading to job stress and changes of institutions. As a result, the public testing phase of the project was pushed back and eventually had to be adjusted to take place during the first lockdown period of the COVID-19 pandemic. Because her IDN work was central to the project, without it the team could not move forward, and long silences from her led to decreased participation. The pandemic also directly affected Dr Hayhurst's ability to participate and communicate on the project, as her microbiology research had direct applications to COVID-19 testing, and thus took her focus entirely (and justifiably) for that time period. All team members were affected by COVID-19, at least indirectly, as it prevented You and CO_2 from going into schools and from planning implementation with teachers busy shifting to online and then hybrid teaching, and as team members found themselves working from home with children and other care responsibilities while also shifting their teaching.

Other tools that enhanced interdisciplinary working on the teams are things that enabled working across institutions. Within institutions, each university uses a system of messaging, video conferencing and file sharing to allow teams to work together; when working with team members in different institutions, however, none of these solutions are applicable, as team members cannot access them. Thus open access and free tools are crucial to interinstitutional team success. Often this entailed some team members learning new systems, as their institutions or disciplines have never made use of them. We trialed a project management app (Asana) on Infectious Storytelling, but found that adding 'just one more thing' to check for the basis of one project too much, and resorted to email. Dr Skains' institution for the IS focus groups insisted on their video conferencing software (Skype for Business), yet its member/ corporate paywall caused it to fail entirely for this purpose, necessitating a move to Zoom and (regular) Skype. We found Google Docs to be the apex of document collaboration, as it enabled real-time co-working, in-document discussion, version history and citation integration with the Zotero reference manager. Likewise, for non-confidential documentation and planning, we have used Google Drive for file-sharing. All of these solutions, like any digital communication and file-sharing, carry risks to privacy and security. Yet interdisciplinary projects across institutions struggle to coordinate themselves without them, and certainly can't make use of single-institution systems.

Moving forward, the key aspect to the continued success of both these projects will be in reviving the personal investment in the research objectives, and in maintaining clear and positive communication with all team members. PI security and stability will hopefully aid in both these areas, as these enable the key project manager to communicate frequently, and to lead on joint tasks like funding applications, publication composition and public impact activities. As long as all team members continue to have shared values, to respect and learn from one another, these interdisciplinary projects should be able to manage future challenges as they have past ones.

6

LESSONS LEARNED: RESEARCHER REFLECTIONS

INTRODUCTION

This chapter is a deliberate attempt to add to the few accounts of wide inter-disciplinarity available (as noted by Klein, 1990), as it sets out reflections from each researcher on the You and CO_2 and Infectious Storytelling projects to date.[1] These reflections were written after the pilot phase of both projects, but before we conducted this more in-depth examination of interdisciplinary studies liter-ature. Interestingly, taken as a corpus, these reflections convey the themes—the benefits and challenges—of interdisciplinary research as outlined in the previous chapter: gaps between silos, distinct disciplinary research cultures and methods, researcher insecurities and biases about interdisciplinary work, the importance of an open approach and appropriate communication and, most crucially, the overwhelmingly positive impact working on these projects has had on our lives.

Every commentary touched in some way on their disciplinary background; these comments illustrate the structural silos almost all researchers become bounded by. Dr Horry summarised her disciplinary silo as

I go to conferences full of cognitive psychologists; I follow other cognitive psychologists on twitter; I read journals full of cognitive psychology research. If I'm feeling really adventurous, I might work with psychologists from other sub-fields.

---

1 Only one reflective commentary is missing, that of Dr Emma Hayhurst. Dr Hayhurst, a molecular biologist at the University of South Wales, rapidly shifted her research team's development of a urinary tract infection test into a rapid test for COVID-19 in Spring 2020 (Thomas, 2020). As a result of this urgent work, she was unable to complete her commentary prior to submission.

Swap out 'cognitive psychology' for almost any discipline, and researcher experiences will be the same. Drs Rudd, Woodward and Casaliggi all describe something similar. The two researchers on the team with previous interdisciplinary experience, Drs Ross and Skains, have perhaps a bit more emotion in their discussions of disciplinarity, with Ross describing a perpetual state of bemusement and discovery, while frustration underlies Skains' note that 'the siloed structures of current HE widen the gap between us—science doesn't understand humanities, humanities doesn't understand science, and there are few opportunities to build bridges'.

Further illustration of research silos, and the challenges in bridging them, is found in the many comments about disciplinary dialects, methods, cultures and metrics of success. These projects highlighted the fact that, as Dr Horry notes, our 'usual interactions with colleagues from [our] own discipline[s] rely heavily on a shared knowledge base'. For these projects to succeed, we have all had to 'learn a whole new set of vocabulary' (Ross), spending 'a lot of time on google learning new words' (Rudd). This process also brought our ways of working into vivid relief, as it 'forced us, in the gentlest of ways to rethink how, what and why we did it' (Woodward), not only with our co-investigators in different disciplines, but with our colleagues within our own areas. Dr Rudd found this aspect a particular challenge: 'It's been a nightmare trying to get my [institutional] grant-writing team to understand that [the You and CO_2 project] is RESEARCH and not outreach' (emphasis original). Reconciling varying methods of collecting data, different forms of data and distinct publishing pathways and metrics is a challenge in both directions: for disciplinarians bridging silos to work in interdisciplinary teams, but also for interdisciplinary researchers and projects trying to communicate and advance their research and their careers within their 'home' disciplinary environments.

It's interesting to look at each of these commentaries in conjunction with their authors' areas of expertise and experience, and what their structure, language and formatting tell us about disciplinary culture. When the lead author (Skains) requested the commentaries, she gave an open remit: she sent a summary of the chapter's objectives as outlined in the book proposal (a bullet point list of six areas of insight about how we worked together). She also sent a series of prompter questions that each co-investigator could use or ignore at their discretion; these questions ask about challenges and positives, lessons learned and examples of experience (most of them are reproduced in Dr Rudd's commentary). Each response demonstrates its author's disciplinary training and culture. Drs Woodward and Casaliggi, the 'purest' humanists in the group, each returned a neatly structured essay, double-spaced in traditional literary manuscript formatting, using formal yet approachable language in describing their experiences. They did not struggle at all to put these commentaries together in what is, essentially, a personal essay form, which reflects the humanities culture and emphasis on the human experience and interaction. Dr Ross, as both a sociologist and a teacher, similarly had no difficulty relaying her perspective in a personal reflection, and her more personable and informal style highlights the openness and friendly approach embraced by educational fields, particularly those who work with young people. Dr Horry, on the other hand, found herself in rather unknown waters, unsure how to frame or format her reflections, as her disciplinary frame of reference in cognitive psychology did not encompass such subjective, discourse-based accounts; her familiar dialect was that of hypothesis, method, results, discussion, quantifiable data and statistical analysis. As a result, she used some of the others' commentaries as a template to construct an essay in a form she'd never experienced before. Dr Rudd, alternatively, eschewed the essay format entirely, treating the question prompts as a pseudo-interview in writing. This is reflective of her very straightforward approach to questions and seeking out their answers; chemistry is a discipline of equations, equal and opposite reactions, and highly structured protocols. It is therefore no surprise that she approached the reflective task as a series of procedures to be completed. Finally, on the other end of the spectrum, Dr Skains, the creative writer, produced a much more personal, emotive account full of illustrative examples, exposition and even fantastical extrapolations of disciplinary origins.

The metaphor of 'bridging silos' is particularly apt when examining the caution researchers use in approaching interdisciplinary work: bridges can be dangerous. Falling between silo gaps—or even not falling—can leave a researcher stranded. Dr Rudd lists the difficulties of explaining her interdisciplinary project to disciplinary colleagues, applying for funding and attempting to publish as just a few of the dangers present in these crevices. Dr Horry notes that crossing these bridges left her, at times, 'feeling like a fish out of water'; Dr Ross claims she's 'always on the backfoot in that I don't always know the terminology or specifics on a topic'; Dr Rudd cites 'imposter syndrome' as one of the significant challenges; and Dr Woodward's feeling of being humbled by 'working with a scientist whose work strives to improve human life in a very direct way' all express the insecurities we face when striking out to work on interdisciplinary projects outside our own silos. These trepidations are echoed not only in our reflective commentaries but also in our frequent communications with one another: we have a tendency when giving feedback to one another (such as on a paper or grant proposal draft) to couch it in 'forgive me if I'm talking about that which I know not' language. We know that we are in unknown territory, potentially blundering into our co-investigators' areas of expertise. We fear, as many do who define themselves by knowledge, appearing ignorant or mistaken, even though by its very nature interdisciplinary research means that we are ignorant of a great deal having to do with our projects.

These insecurities are bolstered by our often culturally ingrained attitudes and biases about other disciplines. Note how Dr Woodward felt 'humbled' by scientists' direct benefit to improve human life, as though film (her discipline) does not have an extensive effect on human life—with the fresh example of the current pandemic, and the mental health and social connection of humanity greatly reliant upon entertainment media, we argue Dr Woodward's work is no less important than Dr Hayhurst's, in the grand scheme of things. Yet we all battle biases, both ours and those of others. Dr Skains recounts how bias was actually *a part of* the Crucible interdisciplinary training programme, in that the facilitators expected the humanists in the room to need an intellectual safety cord, or 'canary', for their interactions with scientists. On the other side of that divide, Dr Rudd (a chemist), recalls that 'I had quite a lot of scientific snobbery before I started the interdisciplinary project, thinking that science had all the answers and could solve all the problems'. These attitudes clashed, initially, as Dr Skains summarises an exchange on the Crucible programme where a scientist assumed the humanists were there to provide science communication skills, a frequent assumption about arts-science collaborations echoed in Dr Woodward's commentary, and indeed much of the research on wide interdisciplinarity (Kelly, 1996).

In order to surmount these challenges, we all describe how attitude adjustments are key to successful interdisciplinary working. Michael Anbar described the ideal interdisciplinary researcher as an 'adventurer' (1973, in Chettiparamb, 2007): we each speak of our approaches to these projects with a tone of adventuring spirit. Dr Horry writes of 'finding my way through a completely new literature', using the metaphor of a way- or path-finder. Dr Skains found herself 'equal measures translator, facilitator, trainer, editor, adviser, and student', reflecting both her roles as one of the few co-investigators coming into these projects with an existing interdisciplinary background, and as a fellow adventurer or 'student' learning new lessons from her colleagues. Dr Ross stresses that 'being open to new work and perspectives is vital to progressing and developing myself', that external guidance aids internal growth—an element Dr Horry notes is crucial to be open to, both on the giving and receiving ends. Dr Woodward describes her pleasure observing Dr Hayhurst's delight in her first foray into archive research, which is a feeling we all share: it *is* delightful to see others learn about our disciplines and begin to see and understand them the same way we do, at least in some small part.

Dr Casaliggi's goal for more people to know and love Romantic era literature and art is a driving factor behind her participation in these projects, and she took keen enjoyment from every recognition of Keats' work and the overlap of medical history with her own discipline, and how each could inform the other.

To achieve this level of enjoyment in teaching and learning about one another's disciplines, of course, the teams had to establish trust, and maintain that trust through appropriate communication. We were very aware of the importance of communication, and how its failure could impact our teams. During the grant writing process, Dr Skains found that her style of communication—rather direct and without some of the hedging noted above—as well as her usual place in teams as the writer could easily cause upset in team members. Her edits and comments on the draft proposal left one team member feeling slighted, and she strove to apologise while conversing more considerately on future elements. As Dr Casaliggi notes, 'it's not all about the books we read or teach and it's not all about our individual academic expertise but it's communication that makes the difference'. We had to work to bridge not only our disciplinary gaps, our dialect gaps and our methodological gaps but also to get to know one another and adjust accordingly (a process that is easier for some than others!).

As described in Chapter 5, we were aided by the various tools at our disposal, including the chat-like comment capabilities of Google Docs, email, instant messaging and, even before the pandemic made all face-to-face interactions virtual, video conferencing. Despite these tools, however, looking back, our best communication and team-building efforts on both teams took place in person. Perhaps vaguely recognising this in our planning stages, each seed funding grant included funds for us to travel from our different institutions for in-person sessions, including the archival research conducted at the National Library of Wales in Aberystwyth for Infectious Storytelling. These meetings were few—one or two per project, under the initial seed grant—but they were key in that they were the only times we met in person outside the overall Welsh Crucible banner. They cemented our trust in one another, our camaraderie, our shared values, and helped to clear up any miscommunications or misconceptions that had formed through mediated communication channels. This is an important aspect of interdisciplinary working to highlight, particularly given the more recent turn to distance-only communication for a great majority of interinstitutional (and, indeed, institutional) research teams, where in-person communication cannot occur, at least for some time. Future interdisciplinary experiments and studies will need to accommodate the loss of face-to-face interactions and the resultant opportunity to build trust and foster working relationships.

It is perhaps appropriate at this stage to point out something that has likely been obvious throughout all these chapters: all of the co-investigators on these teams are women.[2] Originally, You and CO_2 had a slightly different make-up, in that the education and pedagogy role was filled by a male colleague. His working circumstances shifted soon after the project received seed funding, however, and he stepped down, which led to Dr Ross joining the team. This process was not without its hiccups; the months before the situation was resolved were coloured by lack of communication, miscommunication and misinterpreted communication between him and team members. Many reasons could account for this, including a clash of social and/or disciplinary cultural styles, a clash of personalities or a complex of negative emotions affecting all communicating parties—most likely, the difficulties arose because of a combination of all these factors. Once Dr Ross joined the team, the project culture righted itself, and no further communicative or personality clashes occurred (thus far). It is impossible to say whether or not team members' identified gender affected these working relationships; we would be remiss, however, if we did not acknowledge that sharing so many similar demographic qualities across the teams certainly aids in building trust and communication.

This is also not to say that interdisciplinary teams should be composed of contributors from the same demographic groups; in fact, it is more likely that our similarities weaken our research. Note that, until Dr Ross joined and offered a different perspective on socio-economic cultural factors on reception for the You and CO_2 programme, the team had not considered these aspects, which was a significant oversight. What further perspectives and insights could we have achieved with a more diverse team in terms of gender identification, sexual orientation, race, religion, socio-economic status, age, ability and so forth? There is a burgeoning body of literature, for example, demonstrating that indigenous cultures and traditions offer many strategies for combating climate change; moving forward, we would improve the wider understanding and applicability of our projects by including more diverse team members.

Overall, the teams did work well together, despite some early growing pains. Each co-investigator stresses the benefits to our work, in terms of our insights, our research careers, and often even our personal lives. Two of us, Drs Skains and Rudd, changed our career course to focus on health and science communication; we have each left our previous posts for new ones with

---

2 We would also point out that we are all Western, white, cis-het women of similar age and socio-economic backgrounds, though with four different nationalities and speaking at least four languages between us. These factors likely contribute to the cohesiveness of the teams, as we had very few cultural challenges to overcome.

this trajectory. Dr Skains, though she came to these projects with previous interdisciplinary experience, lists the new experiences she gained: 'my first experience doing archive research, my first time teaching secondary school students, my first time leading focus groups, my first "commissioned" creative work'. Dr Rudd's enthusiasm for her new career course reflects in her comments: 'my colleagues are now my friends and so the most fun bit of the project is all of it!... More importantly, I've found my research topic for the rest of my time in academia'. Drs Horry and Woodward each comment upon how these projects have given them new perspectives on their individual disciplines. Dr Horry notes how 'my experiences of working on an interdisciplinary project have opened me up to new ways of thinking and working, and have provided me with new tools that have made me a more rounded researcher', and Dr Woodward echoes this sentiment in writing '[i]t forced me to question what I thought were the limitations of my own field, and signified the potential for an impactful role for film studies in contemporary global challenges'. Likewise, Drs Ross and Casaliggi express excitement at discovering how their disciplines overlap with others, and how these intersections can lead them to new disciplinary insights: 'I've learned things that I can take over into my practice at school' (Ross); 'Although I remain a scholar of Romanticism, this project has taught me that there is value in transdisciplinary research and in using Romantic literature in applied settings, including issues of health and wellbeing' (Casaliggi). Whether we have made significant career changes or simply enjoyed the expansion and intersection of our disciplines with others, we have all gained a wealth of experiences, colleagues, friends and benefits from conducting these interdisciplinary studies.

RESEARCHER REFLECTIONS

This section sets out the reflections for each researcher on the case study projects, with the exception of Dr Emma Hayhurst, who (as noted previously) was unable to provide a commentary due to her necessary work on COVID-19 testing. We have provided the full unabridged commentary for each, to enable others to fully appreciate, review and analyse them for their interdisciplinary insights, as we have been able to do in the preceding section. The formatting of publication will not preserve the subtle differences in the documents as they were submitted; given that we found these differences indicative of disciplinary conventions and cultures, we have described them in the introductory summary to each. There is no particular order to the presentation of these reflective commentaries, though they are grouped by project, with You and CO_2

researchers first and Infectious Storytelling following, with Dr Skains' commentary 'bridging' between the two as the only member of both teams.

Dr Jennifer Rudd

Project: You and CO$_2$
Role: PI, chemistry pedagogy; pedagogical design
Discipline: Chemistry
Commentary formatting: Single-spaced, flush left. Questions presented as bullet points, followed by block answers. No title offered other than the filename. Font: Segoe UI.

How has your conception of other research fields changed as a result of these collaborations?

I've started to appreciate just how important social sciences are. I had quite a lot of scientific snobbery before I started the interdisciplinary project, thinking that science had all the answers and could solve all the problems. The more that I've learnt about climate change from the news and from the project I've realised that actually unless we can communicate our findings effectively then we can have all the solutions in the world, they just won't be implemented!

You and CO$_2$ has been all about communication. Who do we want to communicate with? How will they best listen? How do we deliver information without causing them a whole heap of climate anxiety? How do we impart this information in such a way that the students not only engage in the classroom but also take the message home. Without the amazing *NW4T* story which allows students to explore for themselves, without our psychology expert who can tell us whether we've changed students' attitudes, without our educational expert who thinks about learners who need a bit more help, I wouldn't have been able to tailor the programme effectively. Climate change isn't a technological problem. The solutions are all there. It's now a behavioural problem, an economic problem. For that we don't need science. We need social science.

What have you learned from fields outside your own that you can use to improve your research?

I've learnt that I need to change field. I was a full-time chemistry researcher figuring out how to turn carbon dioxide into a useful product. Whilst this is important work, I knew that I was more interested in taking complex information and communicating it to the general public. This project, and working with my interdisciplinary colleagues, has given me the confidence and the motivation to make the leap away from lab-based chemistry and into the world of science communication.

What were the biggest challenges to working on these projects?

Definitely the language barriers. I spend a lot of time on google learning new words. Hegemonic, Bourdieu, blended learning, interactive digital narrative, habitus…all these words were like a foreign language to me two years ago. Now I can actually not only string a sentence together using my new vocabulary but I can also explain both the qualitative and quantitative components of our project. It was a bit of a shock the first time that I did it!

I think the other challenge can be the imposter syndrome. I work with experts in their fields and as I am a beginner in each of these fields it can be easy to feel unqualified and like I'm not offering anything to the project (And I'm the PI!). I also had to learn that I didn't need to be an expert in each field, I just needed to trust that my expert colleagues would navigate their own field and I had to fund it, organise it and contribute my expertise—chemistry and climate change—wherever it was required. I've also had to overcome being the youngest and least experienced member of the team. My colleagues are senior lecturers and I'm still a post-doc. The only way to overcome that was having a great team. My colleagues chose me as PI and whenever I've lost confidence that is what I've gone back to…they chose me!

Have you found differences in the way different disciplines work, from funding to publishing? What are they?

Oh crikey. It's been a nightmare trying to get my grant-writing team to understand that what I do is RESEARCH and not outreach. When I approached them at the start of the project about further funding they sent me adverts for a little bit of outreach funding. As the project has continued and I've been able to better articulate what it's about and HOW it is research they've been able to assist in finding further funding opportunities a bit better. They've also been won over that this is research!

Publishing has been different as well. For creative writing the onus is on books and IDNs. For my psychology colleague and I multi-authored papers (and preferably many of them) are the norm. For another colleague one, single-authored education paper would be enough per year. In addition, we have to navigate journals that will take interdisciplinary research topics. We've also had funding issues, partly because of the early stage the project is at, but also because the project isn't solely educational, solely about creative writing, solely science but instead an unusual mix of them all. Getting funding bodies to understand that is a challenge.

What are the communication hurdles you've encountered on these projects, and how have you overcome them?

See language above!

What was the most exciting/fun thing you've done as a result of these projects?

It sounds really sappy but my colleagues are now my friends and so the most fun bit of the project is all of it! Even the hard bits aren't so hard because I know my colleagues will cheer me up, and vice versa. Team meetings are a right laugh, although Dr Horry always brings us back to Earth and is key for picking up any flaws in our research plans. More importantly, I've found my research topic for the rest of my time in academia. I can't imagine life without You and CO_2 and I'm really excited about where it will go next. The best thing about interdisciplinary work is that everyone has different ideas for where to take it. I want to add a new workshop on Government agency, our educational expert wants to look at how we could better work with students with special educational needs, our digital expert wants to gameify it further by making an app. There's enough work here for a lifetime and I genuinely feel that I could make a difference, not only to individuals but potentially to an entire generation, if we can embed You and CO_2 into the Welsh Curriculum.

How has working on these projects affected the focus of your research?

Life changing—see above

And of course, anything else you want to include that isn't covered here!
Insights from scientists on communicating their work effectively to non-experts.

I think working on an interdisciplinary team has really helped me to do this. If I'm going to give a science talk now I don't think about the general public as one huge lump but instead think about it more as publics and try to frame my talk accordingly. For example, I was invited to give a talk at a community centre in a low-income, low-level education community. I realised that giving an academic talk would not be correct in this context so I didn't prepare slides and I took my 4-year-old with me. The talk was more of a discussion and the audience appreciated this unusual approach. In contrast, I was invited to give a talk at the Bath Royal Literary and Science Institute, the audience was going to be mostly white collar men in their 60s. There I gave a high-level academic-style presentation. Whilst this might seem logical even without working in an interdisciplinary environment, it was because of the social science aspect of the YCO_2 project that I understood just HOW IMPORTANT it was to meet people where they're at, rather than just giving a talk and expecting people to get the take-home message on their own.

Insights from interdisciplinary project leaders on managing highly disparate disciplinary teams.

Haha I think I might have covered this above, particularly with the imposter syndrome bit.

I do think I've been fortunate that Ruth has a science background as psychology is certainly surprisingly similar to chemistry in that there's rigorous testing, laboratories, control trials and black-and-white answers. In addition, Lyle and Helen both have science/engineering undergraduate degrees so they at least understand my world even if I don't always understand theirs. I've had to learn a new language and my colleagues have had to learn to be patient with me as I try to catch up with their level of knowledge. Grace has been vital in making sure that this project is a success. Overall though it has taught me to really value individual contributions and has helped me understand that to solve the world's major problems we really do need everyone. Each of my colleagues' research and their personal backgrounds brings a new dimension to the project, each time it changes the project and makes it better. As we sit in a time of global chaos with climate change, the pandemic and Black Lives Matter all making us uncomfortable, it is time to embrace our diversity and allow it to bring society forward.

Insights from researchers about funding and publishing interdisciplinary research

Covered this above.

Insights from arts and humanities researchers about their work's role in global challenges.

Insights on research methods employed by different disciplinary fields.

Not much to say here other than it's been great to incorporate quant and qual and see how the project can progress to tie the two together.

Insights on inter-institutional collaboration between researchers.

Not been a problem. Yay for Skype!

Dr Ruth Horry

Project: You and CO$_2$
Role: Quantitative analysis of participant habits and attitudes; design, administer, analyse participant surveys; psychological research
Discipline: Cognitive psychology
Commentary formatting: Double-spaced, first-line indent. No title offered other than the filename. Font: Times New Roman.

I am a cognitive psychologist. My research focuses on how people think and behave. Mostly, I work with other cognitive psychologists. I go to conferences full of cognitive psychologists; I follow other cognitive psychologists on twitter; I read journals full of cognitive psychology research. If I'm feeling *really* adventurous, I might work with psychologists from other sub-fields—social psychologists, perhaps. The You and CO_2 project is my first real foray into truly interdisciplinary research. At times, it has left me feeling like a fish out of water, while at other times, it has pushed me and inspired me to broaden my horizons, and to become a better communicator. Below, I reflect on some of the challenges and delights that I have stumbled across through this project.

Early on, I realised that my usual interactions with colleagues from my own discipline rely heavily on a shared knowledge base. We unthinkingly use technical terms to discuss research design and measurement as we assume that our colleagues will understand them. As a psychologist, I conduct research with human research participants, which brings with it a whole raft of research considerations, each of which has technical language associated with it. When discussing research designs with colleagues from other disciplines, there can be some quite large holes in that shared understanding. Sometimes it's a language barrier—the concept is shared, but we use different terminology to describe it. But at other times, there are concepts that are new to my colleagues, and so I have to take a few steps back and work out how to communicate that concept clearly, and without the technical jargon that is such a part of my usual communication that I no longer recognise it as jargon.

This process absolutely works both ways. I have learned a great deal about qualitative research designs and analysis from my colleagues, who have had to patiently guide me through a new world of sometimes bewildering terminology and ideas. I have learnt to think more deeply about how knowledge is socially constructed by individuals who exist within a broader cultural context, and about the barriers to engagement that can emerge from a misalignment between an individual's cultural context and a topic of discussion.

Another significant challenge for me has been in finding my way through a completely new literature. Who are the key figures in the area? What journals should I focus on? What academic databases should I use and what terms should I search for? At times, it can feel like stumbling around in the dark, without clear direction. It can feel like an insurmountable challenge. But then, you stumble across a term used in a paper that opens up a whole new world—a label for the concept you have been trying to express. And suddenly, a new world of research is opened up to you. In this way, progress comes in fits and starts.

It is difficult to know what you don't know. There can be significant gaps in your knowledge when you venture into a new area, which can be difficult to fill in without someone to guide and mentor you. In my experience, good peer reviewers can serve as useful guides, as they rarely fail to point out where your theoretical argument or background information is lacking. But through this process, you can find yourself exposed to theoretical perspectives that fundamentally change your perspectives and drive the project forward in new, and more meaningful directions.

One of the ways in which my colleagues have broadened my horizons is in thinking about what a research output is. In psychology, the journal article is very much the currency of academic communication. Bluntly put, nothing else really matters. In other disciplines, conference presentations, monographs, technical reports and works of fiction take much more prominent roles. Something that I would previously have considered as a means to an end (e.g. a piece of fiction as a step on the way to a peer-reviewed scientific paper), I can now see as an end in itself.

By and large, I still consider myself to be a cognitive psychologist, and my main research focus hasn't really shifted. However, my experiences of working on an interdisciplinary project have opened me up to new ways of thinking and working, and have provided me with new tools that have made me a more rounded researcher. The lessons I have learned will stay with me for the rest of my research career.

Dr Helen Ross

Project: You and CO_2
Role: Qualitative analysis of participant-submitted IDNs; pedagogical consulting; special educational needs advisory
Discipline(s): Education and sociology
Commentary formatting: 1.08 linespacing, 8pt after-paragraph spacing, flush left. No title offered other than the filename. Font: Calibri.

I have a funny old workload. For my main stay, I teach two days a week in a mainstream secondary school and it's fab. I love my work and it's really good. My teaching role links directly to my research and activism, which is exactly what I was aiming for in doing my PhD. So I was always caught between a couple of fields: education and sociology. I like to think that I've sort of 'officialised' myself in being an educational sociologist with a Bourdieusien slant.

Working across a couple of disciplines and fields isn't without its challenges in my regular life but I love it because of those challenges. I'm always on the backfoot in that I don't always know the terminology or specifics on a topic but I am always happy to admit that. I don't think any learning happens properly under duress so being open to new work and perspectives is vital to progressing and developing myself. Taking part in You and CO_2 has been a sort of baptism of fire into new everythings. I don't know much about chemistry; I stopped at GCSE and did bits and bobs through my engineering degree but it wasn't my favourite thing. I've never studied psychology (I can barely spell it with my dyslexia brain) and I'd not a clue what Twine was before starting to play with it on this project. And I have learned so much.

I've learned about Climate Change Education. It's been amazing. I've learned things that I can take over into my practice at school, which I have done and it's been great with Year 8 during Lockdown. I've learned a whole new set of vocabulary and that bunchems are a thing! I love them. I've seen someone manage a team beautifully, with grace, compassion and efficacy. I've been shown humility when my little boy was poorly and kindness when I've had a rubbish day at work.

Sometimes, the acronyms and terminology associated with teaching and education get a bit much and it can be like my dog has stamped on my keyboard. I've been (rightly) taken to task and challenged with using too many of them and writing like a bit of a wally at times! Use 3 words when you need 3, don't use 14! That was really helpful when writing up a paper. I'm a really good thinker and have some really creative ideas around analysis. Because I am an active classroom teacher, I think I'm quite creative at seeing connections, theorising around them and linking them to the real world of policy and practice. But sometimes those ideas can be foggy because I don't express them well. This project team is so supportive and gracious with critique though. Again, that's just amazing and such a privilege and something to take forward in my own work.

Funding applications are something that I'm not implicated in too much yet as a starting-out independent scholar. I've set up a consultancy to do research, evaluation and consultancy, which runs alongside Special Educational Needs work. I earn money that way and do some research as part of it. But working on the You and CO_2 project, we've put together some funding bids. I'd never seen what they look like or understood the process involved but seeing it, has really given me an invaluable insight for future attempts I may make. I have found a few pots that I may apply for when I've got less on. Work got all busy in the 18 months since I've started out and gone indie so I've not got capacity yet. When I have, I will certainly use the principles and thought processes I've seen on this project.

And the theory! I love the theory! My usual work is based in Bourdieu and I love reading it. I've got a quirky take on some of it and I've loved applying it to different scenarios. I've not used it explicitly for this project as a framework, but my analysis has been gently influenced by my own sociological projects and frameworks. I've read so much around Bourdieu and more on social class.

As climate change education is very associated with certain values, I've read more into Bourdieu's work on values and I will be incorporating that into my further work. It's utterly fascinating. I love it. I always forget how much I LOVE theory and knitting seemingly disparate ideas into something new and applicable in lots of places. This I will take forward and tie to another project that I've got brewing at the moment!

I've loved working on You and CO_2 and it's been such a privilege. I'm proud to be part of such a fab team working on something so important and outside of my usual frame of reference.

Dr Lyle Skains

Project: You and CO_2
Role: Design of project IDN; practice-based research; pedagogical design

Project: Infectious Storytelling
Role: PI; design of project IDN; design, administration, and analysis of focus groups
Discipline(s): Digital creative writing; health and science communication
Commentary formatting: Single-spaced, first-line indent. Title: "Lyle's reflective section – chapter 6". Font: Times New Roman.

I have been in several institutions and through many multi-disciplinary training seminars and workshops, and one thing has always frustrated me: everything about research is set up for scientists. Funding bodies assume you'll need equipment and space and postdocs and supplies; if all you really need is a room of one's own, it can be hard to get funding (which knocks on to promotion and prestige, etc.). University structures assume you'll have a lab full of students writing papers that, of course, you'll put your name in as lab leader. They assume a thesis will be of a certain shape and structure, and are unprepared when it includes a performance or an object. Humanities and arts research is often the red-headed stepchild of academia. Worse, the siloed structures of current HE widens the gap between us—science doesn't understand humanities, humanities doesn't understand science, and there are few opportunities to build bridges.

During the first day of the Welsh Crucible researcher training program (wherein I became involved with all of these brilliantly capable colleagues), our inimitable facilitator asked for a 'humanities canary'. The assumption was that in a room with 30 researchers, most of whom resided in the sciences, the handful of humanities/arts researchers would need a canary who would sing when they didn't understand what the heck a scientist was saying. She initially asked me to do it, but I turned it down, as my first career was as a scientist: I would not be a useful canary. In the end, we humanities researchers were just fine; it was the scientists who frequently needed to sing out that they didn't understand what we were talking about (and fair enough—we throw out concepts like the 'long nineteenth century' and 'hegemonic mores' as often as they toss around *p*-values and ecological diversity).

This moment is a telling one for my experiences on these projects. As someone who has a background in both science and arts research—not to mention professional experience in project management—I found myself equal measures translator, facilitator, trainer, editor, adviser and student. I knew more about infectious disease than about Romantic-era poetry. I have frequently found myself researching and explaining different research methods, publication practices, funding paradigms, teaching approaches and communication strategies, all of which can vary wildly between disciplines.[3] It isn't often that the various research silos mix and mingle (hat's off to the Crucible programme for making that happen); less often do we get to do so beyond a 'wow, you use gold to cure cancer!' or 'I didn't know you could do research in dramatic performance'.

When I first met one of my collaborators, she exclaimed 'Great, you can write a game about my science stuff!'; as my eyes crossed and my heart sunk, a Crucible administrator swooped in to explain that the humanities folks are researchers, too, not just dogsbody communicators.[4] I am grateful every day that she took his words as gospel, and spent quite a lot of time actually *talking* to me about things we were *both* passionate about; we strolled about a lawn on a rare Welsh sunny day on one of our breaks, counting fat furry bee butts and coming up with ideas for how we could change the world using our extremely diverse strengths. To me, this epitomises the evolution of interdisciplinary research from lip service to fully collaborative and ground-breaking work.

These projects have brought me several firsts: my first experience doing archive research, my first time teaching secondary school students, my first time leading focus groups, my first 'commissioned' creative work. We have had our challenges, as any group does, but we have been able to treat them as learning opportunities rather than reasons to compete or hold grudges or step away from the group. The projects have led to career shifts for at least two of

---

3 Chemists are from Mars, mathematicians are from Vulcan, biologists are from Ankh-Morpork, creative writers are from Narnia, film scholars are from Færie, literary scholars are from Queen Elizabeth I's court, and the computer scientists have not yet created a babelfish so we can all reach an accord.

4 Not an actual quote.

us, as we discovered our strengths and enthusiasm for new avenues of research. Not unexpectedly, these shifts have been toward the gaps between research specialisms, seeking to build bridges between different sectors, and to create more space and understanding for interdisciplinary undertakings. If nothing else, that shows how successful these collaborations have been.

Dr Kate Woodward

Project: Infectious Storytelling
Role: Film studies research; analysis of filmic representations of tuberculosis;
analysis of film and animation health communication
Discipline: Film studies
Commentary formatting: Single-spaced, flush left, single line space between
paragraphs. No title offered other than the filename. Font: Calibri.

It was in the South Reading Room of the National Library of Wales that it struck me. There the four of us, researchers from diverse research backgrounds and different geographical areas of Wales, were huddled around a table, enthusiastically digging through the paper archives of the Wales National Memorial Association (Wales' anti-TB programme). Within the sepia coloured paper files neatly bound together with string, as well as policy and strategy relating to attempts to combat the disease, there were personal stories that revealed much about life in Wales at the time. For the three of us who could be loosely categorised as Arts and Humanities researchers, archival work was familiar. But for Dr Emma Hayhurst, the microbiologist of the team, this was a long way away from her lab based work. We watched as she sat delighted, holding the documents, transfixed by the factual reporting of the case of a young girl from the Machynlleth area, whose character and vibrancy some-how bounced off the page. 'This is unbelievable!' she kept exclaiming. 'Welcome to archival research!' was our smiling response.

This anecdote captures the colliding of research backgrounds in the project, which resulted in a questioning of the research paradigms and norms of the field of film studies. The basic differences between us as researchers seemed to amplify themselves in the crucible of the project. From the importance of being the 'first author' on a science paper, to the (for me) natural inclination to write single authored books ('Why would you do that?' was the honest and grounding question from the scientist), the project forced us, in the gentlest of ways to rethink how, what and why we did it.

Aside from the always humbling experience of working with a scientist whose work strives to improve human life in a very direct way, the project highlighted the transformative potential of collaboration across the arts and sciences. The arts and humanities are often seen as a pathway to impact for the sciences, as a vehicle of communicating science to the public. Indeed, utilising cutting-edge digital forms, influenced by findings from traditional arts scholarship, resulted in the potential of a scientific hot topic—that of the very real danger of antibiotics resistance—being propelled and communicated to the public in an impactful way. But its impact went beyond merely *communicating* science, as the project itself created something new; a piece of digital art in its own right, that has its own value, aesthetic and meaning.

Although ultimately we worked within our silos, and didn't leave them, with each of us contributing our own expertise in our usual ways, the project demonstrated the endless possibilities (and fun!) of interdisciplinary work and being made to question, to justify and to learn anew. It forced me to question what I thought were the limitations of my own field, and signified the potential for an impactful role for film studies in contemporary global challenges.

Dr Carmen Casaliggi

Project: Infectious Storytelling
Role: Research into literature and art of the Romantic period/long 19th century; analysis of literature, performance, and artistic representations of tuberculosis
Discipline: English literature (Romanticism)
Commentary formatting: Double-spaced, first-line indent. No title offered other than the filename, but included the proposed chapter title and summary. Font: Times New Roman. Page numbers included, and bracketed word count at the end of commentary.

I am a reader in English literature and the main focus of my research is on Romanticism. I have published widely in my field of expertise: books, edited collections, journal articles, book chapters, book reviews, etc. and all with respectable international publishers. I am also a keen lecturer and I love to infuse my passion for the Romantic period to my students of all ages and backgrounds, both at undergraduate and postgraduate level. Throughout my career I have obtained funding to run international projects in Romantic literature and art. The 'Infectious Storytelling' project is my main/first real interdisciplinary endeavour.

When we obtained funding to run 'Infectious Storytelling' I was very excited. Working with specialists from other disciplines, especially science, meant for me that Romanticism could reach wider audiences if this project proved successful. One of the highlights of this project was to meet with the other Co-Is and the PI in the National Library of Wales in Aberystwyth (May 2019) when we carried out research on the effects and development of TB in our respective disciplines. I was fascinated to find out that sciences and the humanities had so much in common when TB was involved. Not only did I find out that TB shaped the social history of nineteenth-century Europe, but that its impact on the artistic world was just as powerful, with artists offering their own commentaries on the disease through painting, poetry and opera. From John Keats's *Ode to a Nightingale* (1819) and *La Belle Dame Sans Merci* (1819), to Giuseppe Verdi's *La Traviata* (first shown in 1853) and Giacomo Puccini's *La Bohème* (first staged in 1896), consumption was almost a defining feature of the nineteenth century, but was also fundamentally studied in medicine as Robert Koch's pivotal discovery in 1882 for a bacterial cause of the disease shaped modern scientific thought.

Part of our project also involved the organisation and delivery of workshops/focus groups whereby participants were asked to play a digital game based on the 'Infectious Storytelling' project on their phone or laptop. The digital game was created as the result of our individual research in the field of TB. Literature, art, film studies and science shake hands when one starts playing the game. More specifically, one section of the game pays tribute to the poetry on TB of John Keats by referring to his *Ode to a Nightingale*. None of the participants we interviewed understood this Romantic connection and while I was somehow disappointed about this, it also made me realise that perhaps I needed to do more work in order to reach wider audiences with my specific discipline. Similarly, I felt rather bemused when a science teacher thought the reference in the game was to Florence Nightingale or/and to the temporary COVID-19 Nightingale hospital in London. Yet, this taught me that it's not all about the books we read or teach and it's not all about our individual academic expertise but it's communication that makes the difference and if interdisciplinary research serves to deliver Romanticism in new and exciting ways then I'm happy to be involved in this. Nearly every year I show *Educating Rita* (1983) (dir. Lewis Gilbert) to my students as an example of how Romanticism is taught by the protagonist (Michael Caine), a Professor of English Literature, in a University context and I now wonder whether I might try to be even more adventurous in the future and use our interactive digital narrative to gauge students' responses and experience of the period in question.

Furthermore, the focus groups organised around the project taught me that there is room to expand the project more widely while working on other diseases, such as, for example, COVID-19 or/and the Spanish flu of 1918. As a result, I would now like to work more on this project by extending its scope, impactful outcomes and international outlook, and apply for other sources of external funding with some of the scholars (if not all) already engaged in 'Infectious Storytelling' and perhaps even with the involvement of the NHS. I am convinced that there is longevity in the research we have started and that our work is very timely and forward-looking. Although I remain a scholar of Romanticism, this project has taught me that there is value in transdisciplinary research and in using Romantic literature in applied settings, including issues of health and wellbeing.

7

CONCLUSIONS: MOVING FORWARD

If any overarching theme can be identified from the chapters presented in this book, it is *interdisciplinarity*. Here we have presented two distinct (yet overlapping in terms of origins, personnel and methods) interdisciplinary projects using interactive digital narratives (IDNs) for science and health communication (scicomm). Both projects are in the very early stages, and the conclusions presented here offer preliminary feasibility data; it remains to be seen if IDNs as behavioural interventions on issues of global concern—such as climate change and antimicrobial resistance (AMR)—can be said to be significantly effective. Our initial conclusions toward those ends are positive; the works are well received and supported by relevant and robust background research. Yet much more work is needed to definitively state that IDNs for scicomm have an advantage over other methods; this work is planned in the next stages of both You and CO_2 and Infectious Storytelling. What we *can* say that we have learned thus far in these projects is how successful wide interdisciplinary teams, drawing on disciplinary breadth from 'hard' sciences to the creative arts, are fostered and maintained.

The You and CO_2 project discussed in Chapter 2 takes a STEAM approach (recombining Arts practices with STEM education) toward educating young people about climate change. It incorporates chemistry lessons on carbon and carbon footprints, IDN-based bibliotherapy and expressive writing in a series of workshops for secondary school students to encourage them to make positive changes in their lives to reduce their carbon footprints—as well as to take action regarding the overall societal contribution to catastrophic climate change. The pilot phase of this programme demonstrated that it is highly successful in terms of its appeal to teachers and students: engagement from students was strong, and teachers have embraced the programme for in-class use. Our analysis of the pilot surveys and submitted student games, however, has revealed a few gaps in our ability to reach our target audience, primarily in terms of connecting with students whose home and/or cultural environments do not match that of their school or classmates. Future adjustments are underway to better meet the needs of a range of students, including those with special education needs, differing cultural values and English as a second language.

The second of our projects, Infectious Storytelling (Chapter 3), is still in the data analysis phase of its pilot. A great deal of the research for this project was textual analysis—of literature, art, film, games—and the medical history of tuberculosis (TB). The aim of the project was and is to use TB as a case study infectious disease, examining how popular media of each significant period of its history represented the illness, its sufferers and its treatments. We analysed these representations for insights into how popular media can influence attitudes and behaviours affecting public health issues, and these fed into the creation of a purpose-built IDN intended to educate patients about AMR and encourage behaviour changes that can reduce this growing pandemic. Infectious Storytelling has completed its first phase of focus groups, with very positive initial results. The respondents expressed keen interest in the project IDN, indicating they found it compelling and effective as a story, and that its lessons about TB and AMR were surprising, but clear and worth learning. The surprise they expressed is a positive side effect; upon learning something new and surprising, they are thus more likely to convey the information to others via word of mouth, a strong factor in the success of any public health messaging. Like You and CO_2, Infectious Storytelling's next steps will be to revise the IDN to improve upon some technical aspects of the work, better enabling a smooth and enjoyable experience for the player; with further funding and the cooperation of public health bodies, we intend to trial the IDN as public health education and intervention in surgery waiting areas, and for longer-term periods.

You and CO_2 and Infectious Storytelling are both scicomm projects centred on one hypothesis: that using interactive entertainment can successfully effect behaviour change regarding issues of global importance (climate change and AMR, respectively). The key aspects to inspiring behavioural change through fictional narratives is connecting emotionally with the audience, and urging iterative adjustments that the audience believes it can enact (Singhal et al., 2003). Previous successful campaigns in entertainment media have focused on long-running television series; these series have a built-in emotional connection, given their active roles in the target audience's established habits and culture. For new campaigns, particularly those without connections to embedded, long-running series (such as soap operas and radio dramas), the question of emotional connection through narrative, particularly given other practical considerations such as time and budget, becomes much larger. Essentially, the question becomes: how can we emotionally connect with our target audience enough to prompt behaviour change given a very limited budget and likely less than feature-film time frame for our narrative (even less, considering the constraints for Infectious Storytelling)?

While other research teams with different disciplinary constituents might arrive at alternative solutions, for our teams one potential pathway emerged: to connect to our audiences through interactivity and multimodality, perhaps the defining elements of 21st-century media communication. Interactive digital media, in the form of games and narratives from mobile apps to multi-player platforms, are the dominant form of entertainment media in the Western world: with 2.5 billion gamers and a US$151.2 billion market revenue worldwide, the gaming industry dwarfs all other media entertainment industries combined (Dautovic, 2020). Given this global reach and market saturation, it seems prudent to explore how this medium influences player knowledge, perception, attitudes and behaviour. Toward this end, a great deal of pedagogical research attention has examined interactive media and how they can be used to educate and persuade on a variety of topics (e.g. Cheng et al., 2015; de Freitas & Maharg, 2011; Ferrara, 2013; Jarvin, 2015; Mayo, 2009; Ritterfeld & Weber, 2005; Squire, 2011a).

Despite the enormous quantity of interactive works, and the wealth of research that has gone into examining their effects in education and communication, little has been said about specifically *producing* interactive games and narratives for defined educational purposes. Much like creative writing, the body of work done in this area has focused on the texts and their audiences. Numerous guides for game development exist, yet we could find none specifically dedicated to designing interactive stories for education purposes. As discussed in Chapter 4, the shift in rhetorical purpose from 'entertaining any audience' to 'educating a specific audience' is not insignificant. The designer must research and learn about the topic they are communicating and the audience they are communicating to. The creation process is less free-form, with more fixed project goals and pre-determined planning phase. Perhaps most importantly, the creation process is collaborative, though the collaborators are usually not fellow creators; instead, project collaborators in the form of scientists or researchers in other areas more closely resemble clients. Their interest in the project is just as intense as the practitioner's, yet they likely do not understand the creative process or tools. The practitioner is thus required to interpret their colleagues' input, insights and desires for the work, in addition to the other project constraints, which can be both frustrating and fruitful (sometimes simultaneously).

While collaboration on the scicomm IDN can be disruptive, the overall interdisciplinary process is highly rewarding. Both the teams described here benefited from their origins in the Welsh Crucible programme in 2018, which specifically recruited 'adventurous' researcher types open to new ways of working, trained us in interdisciplinary communication and project leadership, and facilitated group formation. The groups thus commenced project activities with a strong foundation of shared values and a sense of fairness and equality regarding each member's discipline and contribution. Our passions for our subject matter—climate change and AMR—drew us together, and our shared goals and our mutual respect of individual objectives fuelled funding applications and research processes. Communication proved essential to the continued development and evolution of both teams, particularly in conveying the purpose and meaning behind discipline-specific dialects, and in being open to receiving these lessons as learners. The trust earned through these exchanges cemented team cultures. As expressed through each team member's reflective commentaries, we have found both career and personal successes and satisfaction through working on these projects.

FUTURE WORK

Given the initial success of both the You and CO_2 and Infectious Storytelling projects in terms of team cohesion and promising pilot results, it is no surprise that both endeavours are moving forward in their research process. You and CO_2 has already secured funding for expansion to more schools using a revised survey, and has adjusted its teacher's pack to allow for online-only education during the COVID-19 pandemic. Several international educational bodies have expressed interest in implementing it in their programmes, which will expand its implementation significantly across nationalities, languages and cultures. This expansion will enable us to dig deeper into the cultural effects of climate change education efficacy, as well as to introduce some longer-term surveys to determine the longevity of the programme's influence.

Infectious Storytelling is moving into phase two of its project plan, as its IDN is revised per focus group feedback. Attention will now turn to renewed funding applications and connections with external stakeholders, primarily public health organisations we can partner with to conduct trials in real-world GP and hospital waiting rooms. This phase of the project has been and may continue to be unfortunately delayed by the COVID-19 pandemic, as most public health bodies are currently focusing the bulk of their efforts on that looming issue. It is conceivable that we may alter our course somewhat, offering our insights into the use of entertainment media for public health education to the more pressing issue of COVID-19 communication, and return to the longer-term challenge of combating AMR once this crisis has passed.

Regardless of the individual pathways and outcomes of these specific projects, we have shown that wide interdisciplinary teams can have great success if provided with a strong foundation in terms of expectations, training and communication. Both project PIs (Drs Jennifer Rudd and Lyle Skains) have altered their career tracks to focus on wide interdisciplinary research: Dr Rudd has gone from a chemistry lab to a management department, leading on a circular economy project that engages researchers and the community in sustainable living. Dr Skains has shifted to a specific 'Health and Science Communication' role where she will be facilitating just the sort of wide interdisciplinary projects we have described here, both internally in her institution and externally. We both draw on the strength of our Crucible training and subsequent experience in order to impart interdisciplinary benefits to more teams and projects.

That is not to say that barriers to interdisciplinary work have been torn down. The most significant barrier to interdisciplinary research—disciplinary silos—remains, from the development of PhD researchers to the measures of academic success. These silos are reinforced by institutional infrastructures such as departments, mailing lists, journal subjects, REF panels and more, which make it difficult to even be aware of, much less connect with, research outside one's own discipline. Increasing pressure on academics' time, as more of their work is casualised and fewer faculty are maintained from year to year even as university management attempts to increase student numbers, leaves researchers with little time to focus on their own disciplinary track, much less to adventure into new territory. Funding bodies, while putting on a good show of emphasising interdisciplinarity in their grant calls, nonetheless struggle to define and understand, and thus *award* interdisciplinary projects. Without funding and support, interdisciplinary teams soon fall back into their familiar, disciplinary patterns of working.

Attempts to tear down these barriers are of course being made, as exemplified in the two new roles Drs Rudd and Skains have moved into. Increasing recognition that science alone cannot explain or influence human behaviour, as Dr Rudd describes in her reflective commentary, is helping to push many efforts at resolving practical, global problems toward interdisciplinary STEAM solutions. Likewise, some institutions are embracing interdisciplinary research and offering internal incentives and funding, enabling more researchers—particularly early career—capacity to develop research centres, training programmes and more. As more individual researchers engage in interdisciplinary teams and projects, perhaps more journals, funders and REF formats will recognise this type of work and help to facilitate it. After all, researchers are *reviewers*; if more researchers acknowledge and embrace interdisciplinary work, they will return more positive reviews of publications, applications and outcomes. We hope that our experiences on these projects, our insights into our working relationships and the demonstration of the benefits of interdisciplinary research on ourselves and our careers show the immense potential for these types of teams and endeavours.

REFERENCES

roseslug. (2018a). r/twinegames - importing data from Google sheets for array in twine (sugarcube 2)? [online] *Reddit*. Retrieved from https://www.reddit.com/r/twinegames/comments/auxx8c/importing_data_from_google_sheets_for_array_in/. Accessed on August 18, 2020.

roseslug. (2018b). Sugarcube 2.21 - SOLVED! Pull in array to game from Google Sheet? [online] *Twine Q&A*. Retrieved from http://twinery.org/questions/46348/sugarcube-2-21-solved-pull-in-array-to-game-from-google-sheet?show=46348#q46348. Accessed on August 18, 2020.

350.org. (2019). 7.6 million people demand action after week of climate strikes. [online] 350.org. Retrieved from https://350.org/7-million-people-demand-action-after-week-of-climate-strikes/. Accessed on March 25, 2020.

Albertalli, B. (2015). *Simon vs. the Homo Sapiens Agenda*. London: Penguin.

Alejo, K. (2016). The CSI effect: Fact or fiction? *Themis: Research Journal of Justice Studies and Forensic Science*, 4, Article 1, 22.

Anbar, M. (1973). The 'bridge scientist' and his role. *Research and Development*, 24(7), 30–34.

Anderson, C. (2006). *The long tail: Why the future of business is selling less of more*. London: Hachette.

Anon. (2013). Welsh crucible. [online] *Welsh Crucible*. Retrieved from http://www.welshcrucible.org.uk/. Accessed on September 8, 2020.

Anon. (2020). Veganuary. [online]. Retrieved from https://uk.veganuary.com/. Accessed on March 25, 2020.

Ardoin, N. M., Bowers, A. W., Roth, N. W., & Holthuis, N. (2018). Environmental education and K-12 student outcomes: A review and analysis of research. *The Journal of Environmental Education*, 49(1), 1–17.

Ashiru-Oredope, D., & Hopkins, S. (2015). Antimicrobial resistance: Moving from professional engagement to public action. *Journal of Antimicrobial Chemotherapy, 70*(11), 2927–2930.

Ashworth, M., White, P., Jongsma, H., Schofield, P., & Armstrong, D. (2016). Antibiotic prescribing and patient satisfaction in primary care in England: Cross-sectional analysis of national patient survey data and prescribing data. *British Journal of General Practice, 66*(642), e40–e46.

Bacigalupi, P. (2010). *Ship breaker*. London: Hachette.

Barker, R. L. (1995). *The social work dictionary*. Washington, DC: National Association of Social Workers, NASW Press.

Barnard, H. (1990). Bourdieu and ethnography: Reflexivity, politics and praxis. In R. Harker, C. Mahar, & C. Wilkes (Eds.), *An introduction to the work of Pierre Bourdieu: The practice of theory* (pp. 58–85). London: Palgrave.

Barry, A., Born, G., & Weszkalnys, G. (2008). Logics of interdisciplinarity. *Economy and Society, 37*(1), 20–49.

Bauer, H. H. (1990). Barriers against interdisciplinarity: Implications for studies of science, technology, and society (STS). *Science, Technology & Human Values, 15*(1), 105–119.

Baum, F. L. (1900). *The wonderful wizard of oz*. Chicago, IL: George M. Hill Company.

Beder, S. (2014). Lobbying, greenwash and deliberate confusion: How vested interests undermine climate change. In M.-C. T. Huang & R. R.-C. Huang (Eds.), *Green thoughts and environmental politics: Green trends and environmental politics* (pp. 18). Taipei: Asia-seok Digital Technology.

Bell, A., Ensslin, A., Ciccoricco, D., Rustad, H., Laccetti, J., & Pressman, J. (2010). A [S]creed for digital fiction. *Electronic Book Review*. [online]. Retrieved from http://www.electronicbookreview.com/thread/electropoetics/DFINative

Berners-Lee, M. (2008). *How bad are bananas?: The carbon footprint of everything*. London: Profile Books Ltd.

Bhattacharya, A., Hopkins, S., Sallis, A., Budd, E. L., & Ashiru-Oredope, D. (2017). A process evaluation of the UK-wide antibiotic guardian campaign: Developing engagement on antimicrobial resistance. *Journal of Public Health, 39*(2), e40–e47.

Blue Planet II. (2017). [Documentary series] *Blue Planet II.* UK: BBC.

Bolton, G., Field, V., & Thompson, K. (2006). Introduction. In G. Bolton, V. Field, & K. Thompson (Eds.), *Writing works: A resource handbook for therapeutic writing workshops and activities* (pp. 13–32). London: Jessica Kingsley Publishers, Ltd.

Bouman, M. (2016). Amusing ourselves to health and happiness: Entertainment media and social change.

Bourdieu, P. (1977). *Outline of a theory of practice.* Cambridge: Cambridge University Press.

Bourdieu, P. (1984). *Distinction: A social critique of the judgement of taste* (Reprint1984 ed. Translated by R. Nice). Cambridge, MA: Harvard University Press.

Boyd, B. (2009). *On the origin of stories: Evolution, cognition, and fiction.* Cambridge, MA: Harvard University Press.

Bracken (Née Bull), L. J., & Oughton, E. A. (2006). 'What do you mean?' the importance of language in developing interdisciplinary research. *Transactions of the Institute of British Geographers, 31*(3), 371–382.

Brandt, D. (1992). The cognitive as the social: An ethnomethodological approach to writing process research. *Written Communication, 9*(3), 315–355.

Brewer, G. D. (1999). The challenges of interdisciplinarity. *Policy Sciences, 32*(4), 327–337.

Brinsley, K., Sinkowitz-CochranR., CardoD., & The CDC Campaign to Prevent Antimicrobial Resistance Team. (2005). An assessment of issues surrounding implementation of the campaign to prevent antimicrobial resistance in healthcare settings. *American Journal of Infection Control, 33*(7), 402–409.

Bromme, R. (2000). Beyond one's own perspective: The psychology of cognitive interdisciplinarity. In P. Weingart & N. Stehr (Eds.), *Practicing interdisciplinarity* (pp. 115–133). Toronto: University of Toronto Press.

Brown, J. S., & Adler, R. P. (2008). Minds on fire: Open education, the long tail, and learning 2.0. *EDUCAUSE Review, 43*(1), 16–32.

Cai, Y., Lu, B., Fan, Z., Indhumathi, C., Lim, K. T., Chan, C. W., Jiang, Y., & Li, L. (2006). Bio-edutainment: Learning life science through X gaming. *Computers & Graphics, 30*(1), 3–9.

Cantell, H., Tolppanen, S., Aarnio-Linnanvuori, E., & Lehtonen, A. (2019). Bicycle model on climate change education: Presenting and evaluating a model. *Environmental Education Research*, 25(5), 717–731.

Cazden, C., Cope, B., Fairclough, N., Gee, J., Kalantzis, M., Kress, G., ... Nakata, N. M. (1996). A pedagogy of multiliteracies: Designing social futures. *Harvard Educational Review*, 66(1), 60.

Chaintarli, K., Ingle, S. M., Bhattacharya, A., Ashiru-Oredope, D., Oliver, I., & Gobin, M. (2016). Impact of a United Kingdom-wide campaign to tackle antimicrobial resistance on self-reported knowledge and behaviour change. *BMC Public Health*, 16(1), 393.

Chambers, A. C., & Macauley, W. R. (2015). Stories about science: Communicating science through entertainment media. [online]. *British Science Association*. Retrieved from https://www.britishscienceassociation.org/blog/stories-about-science-communicating-science-through-entertainment-media. Accessed on August 21, 2020.

Charsky, D. (2010). From edutainment to serious games: A change in the use of game characteristics. *Games and Culture*, 5(2), 177–198.

Cheng, M.-T., Chen, J.-H., Chu, S.-J., & Chen, S.-Y. (2015). The use of serious games in science education: A review of selected empirical research from 2002 to 2013. *Journal of Computers in Education*, 2(3), 353–375.

Chettiparamb, A. (2007). *Interdisciplinarity: A literature review*. Southampton: The Higher Education Academy.

Childerhose, R. K. (1936). Pneumothorax treatment of tuberculosis. *Radiology*, 27(6), 741–748.

Competente, R. J. T. (2019). Pre-service teachers' inclusion of climate change education. *International Journal of Evaluation and Research in Education*, 8(1), 119–126.

Cope, B., & Kalantzis, M. (2009). "Multiliteracies": New Literacies, new learning. *Pedagogies: An International Journal*, 4(3), 164–195.

Cox, D. (2018). Working with Google sheets in twine. *Digital Ephemera*. Retrieved from https://videlais.com/2018/05/16/working-with-google-sheets-in-twine/. Accessed on August 18, 2020.

Csikszentmihalyi, M. (2006). A systems perspective on creativity. In J. Henry (Ed.), *Creative management and development* (3rd ed., pp. 3–17). London: SAGE Publications.

Dahlstrom, M. F. (2014). Using narratives and storytelling to communicate science with nonexpert audiences. *Proceedings of the National Academy of Sciences, 111*(Supplement 4), 13614–13620.

Daniel, T. M. (2000). *Pioneers of medicine and their impact on tuberculosis.* Rochester, NY: University of Rochester Press.

Dashner, J. (2013). *The maze runner.* Frome: Chicken House.

Dautovic, G. (2020). Top video game industry statistics (2020). *Fortunly.* [online]. Retrieved from https://fortunly.com/statistics/video-game-industry-statistics/. Accessed on September 24, 2020.

Dietz, T., Gardner, G. T., Gilligan, J., Stern, P. C., & Vandenbergh, M. P. (2009). Household actions can provide a behavioral wedge to rapidly reduce US carbon emissions. *Proceedings of the National Academy of Sciences, 106*(44), 18452–18456.

Doll, B., & Doll, C. A. (1997). *Bibliotherapy with young people: Librarians and mental health professionals working together.* Westport, CT: Libraries Unlimited.

Donaldson, G. (2015). *Successful futures: Independent review of curriculum and assessment arrangements in Wales.* [online]. OGL. Retrieved from https://dera.ioe.ac.uk/22165/2/150225-successful-futures-en_Redacted.pdf

Dorn, S. (2013). Schools in society. *Educational and Psychological Studies Faculty Publications, 14,* 182.

Drewes, A., Henderson, J., & Mouza, C. (2018). Professional development design considerations in climate change education: Teacher enactment and student learning. *International Journal of Science Education, 40*(1), 67–89.

Dubos, R. J., & Dubos, J. (1987). *The white plague: Tuberculosis, man, and society.* New Brunswick, NJ: Rutgers University Press.

Eatley, G., Hueston, H. H., & Price, K. (2016). A meta-analysis of the CSI effect: The impact of popular media on jurors' perception of forensic evidence. *Politics, Bureaucracy & Justice, 5*(2), 1–10.

Eisenman, A., Rusetski, V., Zohar, Z., Avital, D., & Stolero, J. (2015). Subconscious passive learning of CPR techniques through television medical drama. *Australasian Journal of Paramedicine,* [online] *3*(3), 1–5. Retrieved from https://ajp.paramedics.org/index.php/ajp/article/view/323

Ensslin, A., Skains, L., Riley, S., Haran, J., Mackiewicz, A., & Halliwell, E. (2016). Exploring digital fiction as a tool for teenage body image bibliotherapy. *Digital Creativity, 27*(3), 177–195.

ER. (1994). [TV series] ER. USA: NBC.

Ernst, H., & Colthorpe, K. (2007). The efficacy of interactive lecturing for students with diverse science backgrounds. *Advances in Physiology Education*, *31*(1), 41–44.

Estyn. (2018). *The annual report of her majesty's chief inspector of education and training in Wales 2017–2018*. [online]. Estyn. Retrieved from https://www.estyn.gov.wales/document/annual-report-2017-2018. Accessed on March 25, 2020.

Evans, S. (2019). Analysis: UK's CO2 emissions fell for record sixth consecutive year in 2018. [online] *Carbon Brief*. Retrieved from https://www.carbonbrief.org/analysis-uks-co2-emissions-fell-for-record-sixth-consecutive-year-in-2018. Accessed on September 30, 2020.

Evensen, D. (2019). The rhetorical limitations of the #FridaysForFuture movement. *Nature Climate Change*, *9*(6), 428–430.

Fedunkiw, M. (2003). Malaria films: Motion pictures as a public health tool. *American Journal of Public Health*, *93*(7), 1046–1057.

Ferrara, J. (2013). Games for persuasion: Argumentation, procedurality, and the lie of gamification. *Games and Culture*, *8*(4), 289–304.

Finke, R. A. (1996). Imagery, creativity, and emergent structure. *Consciousness and Cognition*, *5*(3), 381–393.

Flanagan, M. (2009). *Critical play: Radical game design*. Cambridge, MA: MIT Press.

Flower, L., & Hayes, J. R. (1981). A cognitive process theory of writing. *College Composition & Communication*, *32*(4), 365–387.

Frantz, C. M., & Mayer, F. S. (2009). The emergency of climate change: Why are we failing to take action? *Analyses of Social Issues and Public Policy*, *9*(1), 205–222.

de Freitas, S., & Maharg, P. (2011). *Digital games and learning* [online]. London: Bloomsbury Academic.

Garfinkel, H. (1967). *Studies in ethnomethodology*. Englewood Cliffs, NJ: Prentice-Hall.

Gayford, C. (2002). Controversial environmental issues: A case study for the professional development of science teachers. *International Journal of Science Education*, *24*(11), 1191–1200.

Ge, X. (2015). *Emerging technologies for steam education: Full steam ahead* (1st ed.). New York, NY: Springer Science+Business Media.

Gee. (2007). Are video games good for learning? [online].Retrieved from http://cmslive.curriculum.edu.au/leader/default.asp? id=16866&issueID=10696. Accessed on August 28, 2015.

Gibb, N. (2015). The purpose of education. [online]. Retrieved from https://www.gov.uk/government/speeches/the-purpose-of-education. Accessed on August 20, 2020.

Glasemann, M., Kanstrup, A. M., & Ryberg, T. (2010). Making chocolate-covered broccoli: Designing a mobile learning game about food for young people with diabetes. In *Proceedings of the 8th ACM Conference on designing interactive systems, DIS '10* (pp. 262–271). [online] New York, NY: ACM. Retrieved from http://doi.acm.org/10.1145/1858171.1858219. Accessed on July 16, 2019.

Glaser, B. G., & Strauss, A. L. (1967). *The discovery of grounded theory: Strategies for qualitative research*. Chicago, IL: Aldine.

Glaser, B. G. (1978). *Theoretical sensitivity: Advances in the methodology of grounded theory*. Mill Valley, CA: Sociology Press.

Glik, D., Berkanovic, E., Stone, K., Ibarra, L., Jones, M.C., Rosen, B., … Richardes, D. (1998). Health education goes hollywood: Working with prime-time and daytime entertainment television for immunization promotion. *Journal of Health Communication, 3*(3), 263–282.

Goldstein, J. H. (2015). Applied entertainment: Positive uses of entertainment media. In R. Nakatsu, M. Rauterberg, & P. Ciancarini (Eds.), *Handbook of digital games and entertainment technologies* (pp. 1–23) [online]. Singapore: Springer Singapore. Retrieved from http://link.springer.com/10.1007/978-981-4560-52-8_9-1. Accessed on July 18, 2019.

Gombiner, J. (2011). Carbon footprinting the internet. *Consilience, 5*(1), 119–124.

Gordon, G. N. (1839). *Life, letters and journals of lord Byron, with notes* [by T. Moore]. London: J. Murray.

Griffin, D. P. (2017). *CDP carbon majors report 2017* (pp. 16). CDP.

Haviland, S. E., & Clark, H. H. (1974). What's new? Acquiring new information as a process in comprehension. *Journal of Verbal Learning and Behavior, 13*(5), 512–521.

Hayles, N. K. (2002). *Writing machines.* Cambridge, MA: MIT Press.

Heede, R. (2019a). *Carbon majors accounting for carbon and methane emissions 1854–2010 methods & results report.* [online]. Snowmass, CO: Climate Accountability Institute. Retrieved from https://nbn-resolving.org/urn: nbn:de:101:1-2019022204221944769571. Accessed on September 30, 2020.

Heede, R. (2019b). It's time to rein in the fossil fuel giants before their greed chokes the planet. *The Guardian.* [online] October 9. Retrieved from https://www.theguardian.com/commentisfree/2019/oct/09/fossil-fuel-giants-greed-carbon-emissions. Accessed on September 30, 2020.

Heinrick, J. (2006). Everyone's an expert: The CSI effect's negative impact on juries. *The Triple Helix, 3*(1), 59–61.

HESA. (2020). Higher education student statistics: UK, 2018/19 - qualifications achieved. [online] Higher Education Statistics Agency. Retrieved from https://www.hesa.ac.uk/news/16-01-2020/sb255-higher-education-student-statistics/qualifications. Accessed on Jun 26, 2020.

Hether, H. J., Huang, G. C., Beck, V., Murphy, S. T., & Valente, T. W. (2008). Entertainment-Education in a media-saturated environment: Examining the impact of single and multiple exposures to breast cancer storylines on two popular medical dramas. *Journal of Health Communication, 13*(8), 808–823.

Hoffman, S. J., et al. (2017). Celebrities' impact on health-related knowledge, attitudes, behaviors, and status outcomes: Protocol for a systematic review, meta-analysis, and meta-regression analysis. *Systematic Reviews, 6*(1), 13.

Hollingsworth, R., & Hollingsworth, E. J. (2000). Major discoveries and biomedical research organizations: Perspectives on interdisciplinarity, nurturing leadership, and integrated structure and cultures. In P. Weingart & N. Stehr (Eds.), *Practising interdisciplinarity* (pp. 215–244). Toronto: University of Toronto Press.

Huffaker, D. A., & Calvert, S. L. (2003). The new science of learning: Active learning, metacognition, and transfer of knowledge in E-learning applications. *Journal of Educational Computing Research, 29*(3), 325–334.

Hugo, V. (1834). *The hunchback of Notre-Dame.* Carey, ID: Lea and Blanchard.

Idris, F., Hassan, Z., Ya'acob, A., Gill, S. K., & Awal, N. A. M. (2012). The role of education in shaping youth's national identity. *Procedia - Social and Behavioral Sciences, 59*, 443–450.

Inglis, B. (1965). *A history of medicine*. London: Weidenfeld and Nicolson.

Ingram, N. (2011). Within school and beyond the gate: The complexities of being educationally successful and working class. *Sociology, 45*(2), 287–302.

Institute of Medicine. (2015). *Communicating to advance the public's health: Workshop summary*. Washington, DC: The National Academies Press. doi:10.17226/21694

IPCC. (2014). *Climate change 2014: Synthesis report. Contribution of working groups I, II and III to the fifth assessment report of the intergovernmental panel on climate change.* [Core Writing Team, R. K. Pachauri and L. A. Meyer (Eds.)] (pp. 151). Geneva: IPCC.

Ipsos, M. O. R. I. (2020). *GP patient survey: National results and trends.* [online] NHS England. Retrieved from https://gp-patient.co.uk/surveysandreports. Accessed on August 21, 2020.

Izzo, G. M., Langford, B. E., & Vitell, S. (2006). Investigating the efficacy of interactive ethics education: A difference in pedagogical emphasis. *Journal of Marketing Theory and Practice, 14*(3), 239–248.

Jarvin, L. (2015). Edutainment, games, and the future of education in a digital world. *New Directions for Child and Adolescent Development, 2015*(147), 33–40.

Jorgenson, S. N., Stephens, J. C., & White, B. (2019). Environmental education in transition: A critical review of recent research on climate change and energy education. *The Journal of Environmental Education, 50*(3), 160–171.

Karlen, A. (1996). *Plague's progress: A social history of man and disease.* London: Indigo.

Keats, J. (1820). Ode to a nightingale. In *Lamia, Isabella, the eve of St. Agnes, and other poems* (pp. 107–112). London: Taylor and Hessey.

Kelly, J. S. (1996). Wide and narrow interdisciplinarity. *The Journal of General Education, 45*(2), 95–113.

Kesten, J. M., Bhattacharya, A., Ashiru-Oredope, D., Gobin, M., & Audrey, S. (2017). The antibiotic guardian campaign: A qualitative evaluation of an online pledge-based system focused on making better use of antibiotics. *BMC Public Health, 18*(1), 5.

Kim, B., & Neff, R. (2009). Measurement and communication of greenhouse gas emissions from U.S. food consumption via carbon calculators. *Ecological Economics, 69*(1), 186–196.

King, S. (1990). *The stand*. New York, NY: Doubleday.

Kirby, D. A., Chambers, A. C., & Macauley, W. R. (2015). What entertainment can do for science, and vice versa. The Science and Entertainment Laboratory. Retrieved from http://thescienceandentertainmentlab.com/what-ent-can-do-for-sci/. Accessed on August 21, 2020.

Kirby, D. A. (2011). *Lab coats in Hollywood: Science, scientists, and cinema*. Cambridge, MA: MIT Press.

Klein, J. T. (1990). *Interdisciplinarity: History, theory, and practice*. Detroit, MI: Wayne State University Press.

Klein, J. T. (2000). A conceptual vocabulary of interdisciplinary science. In P. Weingart & N. Stehr (Eds.), *Practicing interdisciplinarity* (pp. 3–24). Toronto: University of Toronto Press.

Klimas, C. (2009). Twine. [online] Retrieved from http://twinery.org

Laidley, T. (2013). Climate, class and culture: Political issues as cultural signifiers in the US. *The Sociological Review, 61*(1), 153–171.

Lester, B. T., Ma, L., Lee, O., & Lambert, J. (2006). Social activism in elementary science education: A science, technology, and society approach to teach global warming. *International Journal of Science Education, 28*(4), 315–339.

MacLeod, M. (2018). What makes interdisciplinarity difficult? Some consequences of domain specificity in interdisciplinary practice. *Synthese, 195*(2), 697–720.

Makri, S., & Blandford, A. (2012a). Coming across information serendipitously: Part 1 – A process model [open access version]. *Journal of Documentation, 68*(5), 684–705.

Makri, S., & Blandford, A. (2012b). Coming across information serendipitously: Part 2 - a classification framework [open access version]. *Journal of Documentation, 68*(5), 706–724.

Marzano, M., Carss, D. N., & Bell, S. (2006). Working to make interdisciplinarity work: Investing in communication and interpersonal relationships. *Journal of Agricultural Economics, 57*(2), 185–197.

Mayer, R. E. (2014). *Computer games for learning: An evidence-based approach* [online]. Cambridge, MA: The MIT Press. Retrieved from http://

encore.bangor.ac.uk/iii/encore/record/C__Rb1921087__Sgame%20based%
20learning__P0,5__Orightresult__X1?lang=eng&suite=cobalt. Accessed on
September 1, 2015.

Mayo, M. J. (2009). Video games: A route to large-scale STEM education?.
Science, 323(5910), 79–82.

McKenzie, L. (2015). Narrative, ethnography and class inequality. In
J. Thatcher, N. Ingram, C. Burke, & J. Abrahams (Eds.), *Bourdieu the next
generation: The development of Bourdieu's cultural heritage in contemporary
UK sociology* (pp. 25–36). London: Routledge.

McNeal, P., & Petcovic, H. L. (2019). Sound practices in climate change
education. *Science Scope, 42*(6), 104–107.

Mora, C., Dousset, B., Caldwell, I. R., Powell, F. E., Geronimo, R. C., Bielecki,
C. R., … Trauernicht, C. (2017). Global risk of deadly heat. *Nature Climate
Change, 7*(7), 501–506.

Morens, D. M. (2002). At the deathbed of consumptive art. *Emerging
Infectious Diseases, 8*(11), 1353–1358.

Moulin Rouge. (2001). Directed by B. Luhrmann. 20th Century Fox.

Mugerwa, S., & Holden, J. D. (2012). Writing therapy: A new tool for general
practice? *British Journal of General Practice, 62*(605), 661–663.

NESTA. (2018). NESTA - welcome to crucible in a box. [online]. NESTA.
Retrieved from http://crucibleinabox.nesta.org.uk/. Accessed on September 8,
2020.

New Amsterdam. (2018). [TV series] New Amsterdam. NBC.

Newitt, S., Anthierens, S., Coenen, S., Lo Fo Wong, D., Salvi, C., Puleston, R.,
& Ashiru-Oredope, D. (2018). Expansion of the 'antibiotic guardian' one
health behavioural campaign across Europe to tackle antibiotic resistance:
Pilot phase and analysis of AMR knowledge. *The European Journal of Public
Health, 28*(3), 437–439.

Newman, C. (2020). £3.7m circular economy innovation funding to support
south Wales organisations. [online]. Swansea University. Retrieved from
https://www.swansea.ac.uk/press-office/news-events/news/2020/09/37m-
circular-economy-innovation-funding-to-support-south-wales-
organisations.php. Accessed on September 29, 2020.

Nicieza, F., & Liefeld, R. (1991). *Deadpool.* New York, NY: Marvel Comics.

NIH. (2011). The public health film goes to war: The films. [Digital Library Collections] National Institute of Health. Retrieved from https:// www.nlm.nih.gov/hmd/digicolls/phfgtw/films.html. Accessed on July 5, 2019.

Nissani, M. (1997). Ten cheers for interdisciplinarity: The case for interdisciplinary knowledge and research. *The Social Science Journal*, *34*(2), 201.

Ong, W. (1982). *Orality and literacy: The technologizing of the word*. New York, NY: Routledge.

Outbreak. (1995). Directed by W. Petersen. Warner brothers.

Pardeck, J. T. (2014). *Using books in clinical social work practice: A guide to Bibliotherapy*. New York, NY: Routledge.

Pennebaker, J. W., & Seagal, J. D. (1999). Forming a story: The health benefits of narrative. *Journal of Clinical Psychology*, *55*(10), 1243–1254.

Pennebaker, J. W. (1997). Writing about emotional experiences as a therapeutic process. *Psychological Science*, *8*(3), 162–166.

PERC. (2014). *Blended learning: Defining models and examining conditions to support implementation*. Research for Action. [online] Philadelphia, PA: Philadelphia Education Research Consortium. Retrieved from http:// www.researchforaction.org/wp-content/uploads/2015/11/Blended-Learning-PERC-Research-Brief-September-2014.pdf. Accessed on June 19, 2019.

PHE. (2014). Antibiotic guardian. [online] Public Health England. Retrieved from https://antibioticguardian.com. Accessed on May 11, 2020.

Polivy, J., & Herman, C. P. (2002). Causes of eating disorders. *Annual Review of Psychology*, *53*(1), 187–213.

Pruneau, D., Liboiron, L., Vrain, E., Gravel, H., Bourque, W., & Langis, J. (2001). People's ideas about climate change: A source of inspiration for the creation of educational programs. *Canadian Journal of Environmental Education*, *6*, 121–138.

Pruneau, D., Khattabi, A., & Demers, M. (2010). Challenges and possibilities in climate change education. *US-China Education Review*, *7*(9), 15–24.

Rapeepisarn, K., Wong, K. W., Fung, C. C., & Depickere, A. (n.d.). Similarities and differences between "learn through play" and "edutainment". (p. 5).

Rayner, A. (2019). Labour to make climate change core part of school curriculum. The Labour Party. [online] Retrieved from https://labour.org.uk/

press/labour-make-climate-change-core-part-school-curriculum/. Accessed on March 10, 2020.

Reid, R. W. (1975). *Microbes and men.* New York, NY: Saturday Review Press.

Resnick, M. (2004). *Edutainment? No thanks. I prefer playful learning.* Associazione Civita Report on Edutainment. [online] Retrieved from https://llk.media.mit.edu/papers/edutainment.pdf. Accessed on August 25, 2018.

Ritterfeld, U., & Weber, R. (2005). Video games for entertainment and education. (p. 17).

Robinson, S. K. (2006). TED Talk: Do schools kill creativity? [online] TED. Retrieved from https://www.ted.com/talks/sir_ken_robinson_do_schools_kill_creativity/transcript. Accessed on August 20, 2020.

Rockstar Studios. (2018). Red Dead Redemption II.

Ross, H., Rudd, J. A., Skains, R. L., & Horry, R. (2021). How big is my carbon footprint? Understanding engagement with climate change education. *Sustainability 13*(4), 1961. Accessed February 11, 2021. doi:10.3390/su13041961

Rudd, J. A., Horry, R., & Skains, R. L. (2019). You and CO_2: A public engagement study to engage secondary school students with the issue of climate change. *Journal of Science Education and Technology, 29,* 230–241. [online]. Accessed on March 3, 2020. doi:10.1007/s10956-019-09808-5

Shelley, P. B. (1891). *Adonais.* [online]. London: Clarendon Press. Retrieved from https://www.gutenberg.org/files/10119/10119-h/10119-h.htm. Accessed on June 26, 2020.

Shelton, D. E. (2008). The 'CSI effect': Does it really exist? *National Institute of Justice Journal,* [online] *259,* 1–7. Retrieved from https://papers.ssrn.com/abstract=1163231. Accessed 29 Sep 2020.

Singhal, A., Cody, M. J., Rogers, E. M., & Sabido, M. (Eds.). (2003). *Entertainment-education and social change: History, research, and practice.* Abingdon: Routledge.

Singhal, A. (2013). Introduction: Fairy tales to digital games: The rising tide of entertainment education. *Critical Arts, 27*(1), 1–8.

Skains, L. (2016). The futographer: A hyperstory [hyperfiction]. Retrieved from http://ifdb.tads.org/viewgame?id=3xfnvfag2bqle6ip

Skains, L. (2017). The adaptive process of multimodal composition: How developing tacit knowledge of digital tools affects creative writing. *Computers and Composition*, *43*, 106–117.

Skains, R. L. (2018). Creative practice as research: Discourse on methodology. *Media Practice and Education*, *19*(1), 82–97.

Skains, R. L. (2019a). Teaching digital fiction: Integrating experimental writing and current technologies. *Palgrave Communications*, *5*, 1–10. [online] Retrieved from https://www.nature.com/articles/s41599-019-0223-z

Skains, R. L. (2019b). The materiality of the intangible: Literary metaphor in multimodal texts. *Convergence: The International Journal of Research Into New Media Technologies*, *25*(1), 133–147.

Sood, S., Menard, T., & Witte, K. (2003). The theory behind entertainment-education. In A. Singhal, M. J. Cody, E. M. Rogers, & M. Sabido (Eds.), *Entertainment-education and social change: History, research, and practice* (pp. 117–149). Abingdon: Routledge.

Squire, K. (2011a). *Video games and learning: Teaching and participatory culture in the digital age*. Kurt Squire ; foreword by James Paul Gee ; featuring contributions by Henry Jenkins. [online]. New York, NY: Teachers College Press. Retrieved from http://encore.bangor.ac.uk/iii/encore/record/ C__Rb1902771__Sgame%20based%20learning__P0,1__Orightresult__X3? lang=eng&suite=cobalt. Accessed on September 1, 2015.

Squire, K. (2011b). *Video games and learning: Teaching and participatory culture in the digital age. Technology, education–connections (the TEC series)*. New York, NY: Teachers College Press.

Stewart, J. (2018). Twine game data to google sheets via javascript version 2. John Stewart. Retrieved from https://johnastewart.org/coding/twine-game-data-to-google-sheets-via-javascript-version-2/. Accessed on August 18, 2020.

Stirling, A. (2014). Disciplinary dilemma: Working across research silos is harder than it looks. *Guardian Political Science Blog*. [online]. Accessed on August 21, 2020. doi:10.13140/RG.2.1.1919.3680

Sullivan, G. (2009). Making space: The purpose and place of practice-led research. In H. Smith & R. T. Dean (Eds.), *Practice-led research, research-led practice in the creative arts* (pp. 41–65). Edinburgh: Edinburgh University Press.

Tagore, P. (2000). Keats in an age of consumption: The 'ode to a nightingale'. *Keats-Shelley Journal*, *49*, 67–84.

Taylor, M. (2019). Teachers want climate crisis training, poll shows. *The Guardian*. [online] June 21.Retrieved from https://www.theguardian.com/environment/2019/jun/21/teachers-want-climate-crisis-training-poll-shows. Accessed on March 10, 2020.

Terry, N., Macy, A., Owens, J., & Womble, L. (2016). *Dangerous foods: The libel case of the Texas cattlemen versus oprah winfrey*. [online] London: SAGE Publications. Retrieved from http://sk.sagepub.com/cases/dangerous-foods-libel-case-of-texas-cattlemen-versus-oprah-winfrey. Accessed on August 20, 2020.

The Citadel. (1938). Directed by K. Vidor. MGM.

Thomas, H. (2020). University's rapid test could be used 'in weeks'. *BBC News*. [online]. April 20Retrieved from https://www.bbc.com/news/uk-wales-52347827. Accessed on September 16, 2020.

Tombstone. (1993). Directed by G. P. Cosmatos. Buena vista. Retrieved from https://www.imdb.com/title/tt0108358/. Accessed on June 26, 2020.

Tomlinson, S. (2011). *A sociology of special education*. Oxon: Routledge.

Valente, T. W., & Bharath, U. (1999). An evaluation of the use of drama to communicate HIV/AIDS information. *AIDS Education and Prevention*, 11(3), 203–211.

van der Bom, I., Skains, R. L., Bell, A., & Ensslin, A. (2021). Reading hyperlinks in hypertext fiction: An empirical approach. In A. Bell, S. Browse, A. Gibbons, & D. Peplow (Eds.), *Style and reader response: Minds, media, methods* (pp. 123–142). Amsterdam; Philadelphia, PA: John Benjamins Publishing Co.

Wang, H., & Singhal, A. (2016). East los High: Transmedia edutainment to promote the sexual and reproductive health of young Latina/o Americans. *American Journal of Public Health*, 106(6), 1002–1010.

Ward, T. B., Smith, S. M., & Finke, R. A. (1999). Creative cognition. In R. Sternberg (Ed.), *Handbook of creativity* (pp. 189–212). New York, NY: Cambridge University Press.

Webb, T. L., Sheeran, P., & Luszczynska, A. (2009). Planning to break unwanted habits: Habit strength moderates implementation intention effects on behaviour change. *British Journal of Social Psychology*, 48(3), 507–523.

Wegner, D. M., & Schaefer, D. (1978). The concentration of responsibility: An objective self-awareness analysis of group size effects in helping situations. *Journal of Personality and Social Psychology*, 36(2), 147–155.

Weiler, L. (2018). Making immersive experiences 1. Public Lecture. Manchester Metropolitan University, Manchester, UK, 8 June 2018.

Weingart, P., & Stehr, N. (Eds.). (2000). *Practising interdisciplinarity.* Toronto: University of Toronto Press.

Welsh Government. (2015). The four purposes of the curriculum for Wales. [online] gov.wales. Retrieved from https://gov.wales/sites/default/files/publications/2018-03/the-four-purposes-of-the-curriculum-for-wales.pdf. Accessed on March 25, 2020.

Whittier, D. K., Kennedy, M. G., St. Lawrence, J. S., Seeley, S., & Beck, V. (2005). Embedding health messages into entertainment television: Effect on gay men's response to a syphilis outbreak. *Journal of Health Communication, 10*(3), 251–259.

WHO. (2015). Global action plan on antimicrobial resistance. [online] World Health Organization. Retrieved from http://www.who.int/antimicrobial-resistance/publications/global-action-plan/en/. Accessed on August 20, 2018.

WHO. (2017). *Antimicrobial resistance behaviour change first informal technical consultation.* [online] Château de Penthes, Geneva: World Health Organization. Retrieved from http://www.who.int/antimicrobial-resistance/AMR-Behaviour-Change-Consultation-Report_6-and-7-Nov-2017.pdf. Accessed on August 20, 2018.

WHO. (2018). Drug-resistant tuberculosis. [online] World Health Organization. Retrieved from http://www.who.int/tb/areas-of-work/drug-resistant-tb/en/. Accessed on August 20, 2018.

WHO. (2019). World antibiotic awareness week. [online] World Health Organization. Retrieved from https://www.who.int/news-room/campaigns/world-antibiotic-awareness-week/world-antibiotic-awareness-week-2019/landing. Accessed on May 11, 2020.

Wiedmann, T., & Minx, J. (2007). *A definition of carbon footprint.* Durham: ISA-UK.

Wise, S. B. (2010). Climate change in the classroom: Patterns, motivations, and barriers to instruction among Colorado science teachers. *Journal of Geoscience Education, 58*(5), 297–309.

Wright, R. R., & Sandlin, J. A. (2009). Cult TV, hip hop, shape-shifters, and vampire slayers: A review of the literature at the intersection of adult education and popular culture. *Adult Education Quarterly, 59*(2), 118–141.

INDEX

Printed in the United States
by Baker & Taylor Publisher Services